Voices of the Rocks

A SCIENTIST LOOKS AT CATASTROPHES AND ANCIENT CIVILIZATIONS

ROBERT M. SCHOCH, Ph.D.
with Robert Aquinas McNally

HARMONY BOOKS • NEW YORK

THIS BOOK IS DEDICATED TO

Cynthia Schoch
everything a wife should be

Nicholas Schoch and Edward Schoch
two inspirational fellows

Gayle Eleanor
woman of love and poems

Darren Matthew McNally and Brian David McNally
wild men for wild times

Published by Harmony Books, 201 East 50th Street, New York, New York 10022
Member of the Crown Publishing Group
Random House, Inc. New York, Toronto, London, Sydney, Auckland
www.randomhouse.com

HARMONY BOOKS is a registered trademark and Harmony Books is a trademark of Random House, Inc.

Printed in the United States of America

Library of Congress Cataloging-in-Publication Data

Schoch, Robert M.
Voices of the rocks: a scientist looks at catastrophes and ancient civilizations / Robert M. Schoch with Robert Aquinas McNally.
Includes bibliographical references.
1. Catastrophes (Geology) 2. Civilization, Ancient. 3. Lost continents.
I. McNally, Robert Aquinas. II. Title.
QE506.S36 1999
900—dc21 98-56075

ISBN 0-609-60369-8

10 9 8 7 6 5 4 3 2 1

First Edition

Acknowledgments

MANY PEOPLE HAVE INFLUENCED AND ENCOURAGED ME OVER THE YEARS. I was fortunate to attend George Washington University (Washington, D.C.) as an undergraduate student and Yale University as a graduate student; at both of these institutions I was exposed to many fine faculty members, staff, and students who helped shape my thinking and interests. Since 1984 I have been a full-time faculty member at the College of General Studies, Boston University. The administration, faculty, staff, and students at the College have always been supportive of my studies, and I continually find the interdisciplinary interactions with my colleagues there refreshing and exhilarating. Dr. Brendan Gilbane, Dean of the College, has always been encouraging toward my endeavors, lending advice and support, as did the late Dr. Charles P. Fogg, former chairman of the Division of Science at the College of General Studies.

My research concerning the age of the Great Sphinx would not have been possible without the help and cooperation of many individuals and organizations. In particular, I thank Drs. Mohamed I. Bakr, Ali Hassan, Zahi Hawass, Elsayed Hegazy, and Abdul-Fattah El-Sabbahy (all of the former Egyptian Antiquities Organization) for permission to pursue geological and geophysical studies on the Giza Plateau. Ahmed Nour-el-Din was extremely helpful during my March 1993 trip to Egypt, as was Nehad Gamal. I thank Drs. Gaber Barakat, L. Abdel-Khalek, M. M. El Aref, and Eglal Refai (all of the Faculty of Science, Cairo University) for their interest, advice, and help. Dr. Thomas L. Dobecki (formerly of McBride-Ratcliff and Associates, Houston) provided invaluable assistance with the geophysical studies on the Giza Plateau. The staff at Geosignal, Inc. (Houston), contributed their talents in processing the high-resolution seismic reflection data. Dr. Frank Ward of the American Embassy in Cairo also helped me immensely, giving freely of his time and expertise. Dr. Hanny M. El Zeini provided needed encouragement early in the Sphinx project. I also extend thanks to Drs. Lambert T. Dolphin (formerly of Stanford Research Institute) and Mark Lehner (University of Chicago) for helpful discussions, especially during the early stages of this project, and to Drs. James J. Hurtak and Robert G. McKinney for helpful advice and discussion more generally.

Dr. Robert Eddy (formerly of the College of Basic Studies, renamed the College of General Studies in September 1992, of Boston University) first introduced me to John Anthony West, and it was as a result of discussions with West that I became interested in the problem of the age of the Sphinx. Thus, West is responsible for initiating this aspect of my research, and he and Boris Said deserve credit for their hard work relative to the logistics of the Sphinx project. In particular, Boris undertook most of the hammer-slinging during the April 1991 seismic studies (I will ignore, for the moment, his suggestion that the Sphinx is nothing but a fancy hood ornament on a giant sports car buried under the sand). Here I express my thanks to both West and Said for their continuing help and encouragement.

It was through the generosity of Yasuo Watanabe (Japan Medical Dynamic Marketing [MDM], Inc., Tokyo) that I was able to visit Yonaguni Island and explore the Yonaguni Monument on 23–24 September 1997. I sincerely thank him for his hospitality and patronage. Graham Hancock was instrumental in arranging the trip to Yonaguni, and he and his wife, Santha Faiia, joined me there to dive on the monument. John Anthony West accompanied me to Yonaguni, participated in the scuba diving, and engaged me in many deep discussions (as he always does) concerning the Yonaguni Monument and related matters. Shun Daichi (Tokyo correspondent of *New Perspectives Quarterly*) also joined us in diving, and provided me with various reference materials on the Yonaguni Monument and help in translating. Kihachiro Aratake, who actually discovered the Yonaguni Monument, was our guide for the dives, and dive masters Hiroshi-Kubota and Yoshimi Matsumura provided us with invaluable assistance, as did Megumi Kondo (MDM), Akiko Ito (MDM), Ken Yamada (Ortho Development Corporation, Utah), Dale Kimsey (Ortho), and Steve Hubbard (Ortho). Dr. Michael Pill and Linda Moulton Howe provided me with copies of references from the Japanese *Super Mystery Magazine Mu.*

I returned to Japan in late July and early August 1998 to further study the Yonaguni Monument and related structures as a member of the Team Atlantis multidisciplinary underwater research team and documentary film project. I thank Michael Arbuthnot for organizing this effort and inviting me to participate. Boris Said, producer for Team Atlantis, once again was immensely helpful in making sure that everything worked out. The project would have been impossible without the assistance of Iris DeMauro, who gave freely of her time and energy throughout the expedition and allowed us to stay at her family's beach house in Okinawa during the latter half of the trip. Matthew Sapero, webmaster for the Internet site www.teamatlantis.com, engaged me in much stimulating discussion during the trip and helped me with all things related to computers (including the sending and receiving of E-mail and files so that I could continue working on this book even while in Japan). Dr. J. J. Hurtak and Dr. John T. Dorwin also participated in this trip; sharing freely of their knowledge and opinions, they gave me much food for thought. Also in Japan with the Team Atlantis 1998 expedition were Janet Arbuthnot, Christopher DeFelice, Sarah Kingston (who provided invaluable assistance to me underwater), Peter McDougall, Vince Pace, D. J. Roller, and Sandy Wright. I thank them all for their participation. In Japan we were assisted by many people; in particular I would like to single out Kihachiro Aratake (again, as in 1997, he was invaluable in arranging the diving in Yonaguni), Atsushi Mori (who helped us in many ways), and Chie Mikami of Ryukyu Asahi Broadcasting Corporation (who helped us locate an underwater site in the Chatan area of Okinawa). Finally, I extend heartfelt thanks to Dr. Masaaki Kimura for his hospitality. Dr. Kimura met with us numerous times during our stay and explained in detail his ideas regarding the Yonaguni Monument and similar structures. I greatly appreciate his years of research.

Innumerable other persons have taken an interest in my research more generally, and have helped out in various ways. Unfortunately, it would be virtually impossible to list everyone I have had contact with concerning my research, but I would like to acknowledge the help, support, and interest of a few more people in particular, including: Alessandro Abdo, Sean Adair, Alicia Alexander, Bodhi Annan, Millard Baublitz Jr., Robbin Beauchamp (who carpooled with me to Boston University during the height of the Sphinx controversy, and heard more than she probably cared to about the Sphinx), Paul Bierman, David P. Billington Jr., Peter E. Busher, Sita Chaney, John Cheshire, Molly Cheshire, Rita Corriel, Bill and Carol Cote, Barbara Crane, Anh Crutcher, Nelson Dale, Caroline Davies, William E. Davis, Kate Dickie, Lloyd Dickie, Frank Domingo, Michael Eldredge, Dennis C. Forbes, Tim Friend, Jean A. Garrison, Alison and Gilbert Gjersvik, Linda Goldstein, Diane Graszik, Jim Green, Terence Guardino, Eileen P. Gurly, Foxy Gwyne, Samuel Hammer, Khadija S. Hammond, Marta J. Hersek, John Hipsley, Sandy and Chuck Houghton, Natalia Ignatieff, Nicholas Ignatieff, Jorjana Kellaway, Michael G. Kort, Devendra Lal, Mary Lomando, Nancy Long, Michael G. Mahon, Michele Matthews, John E. Miller, Mohammed Nazmy, Richard W. Noone (who pushed me hard to write this book and supplied me with various useful leads and reference materials), T. C. Onstott, Robert Oresick, Matthew A. Parfitt, Marshall Payn, Cheryl Poppaw, Andris and Margaret Priede, Peyton E. Richter, Paul William Roberts, Joseph Schor, Tamer Shawsky, Diana Shields, C. Simonds, Lydia Smithers, David Solomon, Sally Sommers Smith, Richard Speir, Samuel Stern, Gigi Van Deckter, Robert M. Watts, Celesta West, Joanne Wright, and Stan Zippan. I apologize to anyone I have inadvertently missed.

Finally, I must acknowledge the support that my immediate family has provided as I have pursued my studies (which have sometimes entailed my absence from home for weeks at a time). My wife, Cynthia Pettit Schoch, and my two sons, Nicholas Robert Schoch and Edward Robert Schoch, have always been there when I needed them. Additionally, my parents, Milton and Alicia Schoch; my sister, Marguerita C. Schoch; my late grandmother, Adriana M. Goetz; and my parents-in-law, Robert and Anne Pettit, have always been supportive of all my endeavors.

This book could not have been written without the talent, energy, intelligence, and enthusiasm of my coauthor, Robert Aquinas McNally. I could not have asked for a finer collaborator. I am deeply indebted to him for all his help.

Robert and I also benefited from the excellent work of our two fine literary agents, Sarah Jane Freymann and Judith Riven, who combine a love of good books with a sharp sense of business. From first idea to final production, this book benefited greatly from the good help of the people at Harmony Books: namely, Laura Wood, Patricia Gift, Kristen Wolfe, Brian Belfiglio, Ari Gersen, and David Wade Smith.

Of course, all matters of fact and interpretation expressed in this book are solely my responsibility. If there are mistakes, they're mine.

Robert M. Schoch
College of General Studies
Boston University
August 1998

A Personal Note

THIS BOOK REPRESENTS A PERSONAL INTELLECTUAL JOURNEY. I WAS RAISED a uniformitarian, but through the course of my research I have come to doubt the dogmatism that seems to be central to so much of what currently passes for science. Yet I adamantly remain a scientist committed to the scientific method. The purpose of this book is to question the dominant paradigm and tentatively propose an alternative working model. In no way am I, in turn, dogmatic about the speculations presented in this book. Rather, I hope to encourage the reader to think about, and question, some of the larger issues and themes that face us. If the reader comes away from this book simply looking at the world from a slightly different angle, perhaps even with a slightly different perspective, then I have accomplished the task I set for myself.

Contents

There have been, and will be again, many
destructions of mankind arising out of many causes;
the greatest of these have been brought about by the
agencies of fire and water....

PLATO, *Timaeus*

Introduction

In my imagination I could hear the voice of the rock cutting through the atmosphere to strike the Earth—the dramatic, terrifying scream that eyewitnesses of meteorite impacts report. As space rocks go, this one was of fair to middling size, about two-thirds of a mile across, weighing in the substantial neighborhood of 2 billion tons, and moving at more than 50,000 miles per hour. This substantial mass, traveling at that incredible speed, slammed into dry land with such unimaginable force that it vaporized on impact, releasing energy equivalent to 50,000 to 100,000 megatons* of TNT—far larger than the simultaneous detonation of the contemporary world's entire nuclear arsenal. Most likely, the force blew out the atmosphere over the impact site, sending an enormous, fiery plume of dust and debris well above the atmosphere to darken the sun and dim the sky, possibly for weeks or months, perhaps over much of the globe. All this happened in an instant, and in that same instant, every living organism within hundreds or even thousands of miles of the collision site was killed outright.

Yet, as it will, life returned to this site of complete devastation, now called the Haughton Astrobleme (or impact crater, from

***A megaton equals 1 million tons. Currently the explosive potential of the world arsenal of nuclear weapons is estimated to equal 20,000 megatons of TNT.**

roots meaning literally "star wound") located on Devon Island in the Canadian High Arctic, some 900 miles below the North Pole. The twelve-and-a-half-mile-wide crater left by the impact filled with water to become a lake, and the sediments of that ancient lake transformed the remains of the plants and animals living in and around the water into fossils. It was the fossils that brought me, a newly minted Yale University geologist and paleontologist, to find out more about the story they told.

The world those fossils described, the one that flourished on the order of 20 million years ago, during the early Miocene epoch, was strikingly different from today's Arctic, which is a vast, bleak, demanding landscape. Few mammals can survive here other than the polar bear, Arctic fox, lemming, caribou, and ringed seal, creatures well adapted to the deep cold and sunless dark of the long northern winter. The low temperatures and the seasonality of the light likewise limit the plant community to a few low-growing species that hug the ground as protection against freezing and wind damage. Yet, remarkably, this severe world was once softer, even luxuriant—and not very far in the past, by geological standards.

Twenty million years ago, during the period when the asteroid struck, Devon Island was covered with a forest of birch trees and conifers, a landscape similar to what one now finds about 2,000 miles to the south, in Minnesota, Wisconsin, and Maine. Now-extinct forms of rhinoceros and mouse deer browsed among the trees; shrews and pika-like relatives of modern rabbits darted through the shadows; and freshwater fish swam the lakes and streams.

Even farther back, on the order of 45 to 65 million years ago, during the Paleocene and Eocene epochs, the fossil record shows Devon Island to have been still more profoundly different. Back then, what is now the Arctic was a region of swampy lowlands, slow-moving rivers, and towering forests of dawn redwood, kadsura, and ancestral forms of hickory, elm, birch, sycamore, and maple. Primitive fishes, crocodiles, salamanders, newts, and turtles inhabited the rivers and marshes, while the forests and meadows supported flying lemurs, early primates, forerunners of today's cats and dogs, and ancestors of the rhinos, tapirs, and horses.

As I worked at collecting and cataloging the rich lode of fossils my colleagues and I uncovered in the Haughton Astrobleme, I imagined the terrible drama of the asteroid striking Earth. I found myself wondering, too, about time and change. I had finished my Ph.D. just the year before, and the lessons of graduate school were still fresh in my mind. The physical and biological world, I had been taught, changed only very slowly. Evolution was a gradual process, requiring great lengths of time for each step. Somehow, though, this model didn't fit well with what I was seeing all around me. For one thing, there was nothing gradual about the collision that had formed the Haughton Astrobleme, and for the organisms throughout the surrounding region, the collision was nothing less than the ultimate catastrophe. Likewise, slow change didn't account well for the profound shifts in this patch of Earth over the past 65 million years, particularly during the most recent 20 million years, which is hardly long at all in geological terms. Obviously the Arctic was much warmer then than now, but was it dark for four and a half months? How did plants so similar to those now found in the temperate zones survive the protracted darkness? Or could it be that the land now in the Arctic lay then at a much lower latitude and drifted north over the intervening millions of years? Could the ancient Earth have turned on a different axis, so that yesterday's North Pole was somewhere else, and the Arctic now wasn't the Arctic then?

There was much more to my thinking than idle ivory-tower speculation about the physical and biological nature of a now-vanished Earth. If our planet had undergone major, even sudden changes before humans appeared, then it was probably subject to similar dramatic alterations within the span of our species's experience as well. And, since history is the blueprint of the future, whatever happened then was likely to occur yet again. What has been is the key to what will be. It's something we need to know.

Fortunately, my graduate school education provided a good foundation for such thinking. Even though I learned the prevailing uniformitarian model of slow, gradual change, the field of geology in the 1970s and early 1980s was a hotbed seething with radically new ideas. I was among the first generation of scientists to go into geology following the plate tectonics revolution of the 1960s, and the fast-changing field welcomed revolutionary ways of thinking that could stand the test of scientific scrutiny.

It was this intellectual ferment that first drew me to geology while I was an undergraduate at George Washington University. Originally interested more in physical anthropology and human evolution, I found myself fascinated by geology, not because everything was known, but because so much was unknown. The professor in my introductory class readily admitted that enormous gaps existed in our knowledge of Earth and its processes, and that a great deal remained to be discovered. On field trips, I watched experts from the Smithsonian Institution and the United States Geological Survey disagree openly about the significance of the rock formations we were examining. This excited me. I wasn't looking for pat answers and explanations to be memorized for the next examination. I wanted discovery, insight, and knowledge. Geology offered them all.

I completed bachelor's degrees in both anthropology and geology, then set off for Yale and graduate school in geology. Over the next four years I earned two master's degrees and a Ph.D., focusing on the study of fossil vertebrates and writing my dissertation on an extinct order of mammals. I spent a year working at the Peabody Museum of Natural History, the job I held during my summer's work inside the Haughton Astrobleme. Soon after I returned, I began teaching at Boston University's College of General Studies, where I am now a tenured associate professor.

• • •

My Devon Island speculation on the connections between time, catastrophe, and human history was piqued yet again in late 1989, when a friend at Boston University introduced me to John Anthony West. West is an independent Egyptologist who makes his living by writing about ancient Egypt and leading tours to the country and its monuments. Among academic circles, West has a reputation as a maverick and a gadfly. In talking with him, I soon discovered West to be brilliant and many of his ideas predictably outrageous. Still, one of his notions intrigued me. West was certain that the Great Sphinx of Giza had been built thousands of years earlier than the conventionally accepted date of circa 2500 B.C.

His position intrigued me for two reasons. One was the lure of ancient history, which has interested me since childhood. As a boy of twelve, I discovered a 1,600-year-old Roman coin in a flea market, and the thrill of that find remains a memory as fresh as today. The hours of library research I devoted to dating the coin—it proved to come from the reign of the Emperor Valens, in the late fourth century A.D.—instilled in me a fascination with ancient civilizations, one that West shared. The second reason was the chance to bring my scientific training to bear on a significant historical question. Geological methodology, I realized, could be used to evaluate West's hypothesis on the date of the Sphinx. This wasn't speculation. It was science, the testing of new ideas against the exacting measure of fact.

By the following summer I was in Egypt with West, researching the Sphinx. After two more trips to pursue various studies of the monument and its surroundings, I was sure that West's intuition was correct. The data were absolutely convincing: The oldest portions of the Sphinx date to at least 5000 B.C., and may well be even older.

Over the following months and years I published my findings, presented them to scholarly meetings, and was even interviewed on a few television programs. Yet, despite the overwhelming preponderance of evidence, the conventional Egyptological establishment rejected my older date for the Sphinx. They had a reason, one much larger than simply a vested academic interest in this single ancient monument. The Sphinx is obviously the work of a civilization that was highly advanced artistically and technically. Pushing the date of the Sphinx back to a time when humans are conventionally thought to have been primitive Stone Age villagers poses a huge, troubling question. Who were these unknown people? What became of them and their culture? And how does their existence—and disappearance—alter our view of the course of human history and the origins of civilization?

As I have followed the intellectual trails revealed by the redating of the Sphinx, and talked for long hours with fellow researchers working on parallel lines of inquiry, I have become convinced that we are in the midst of a profound scientific paradigm shift. The predominant notion that our species inhabits a slowly changing, steady-state planet is falling by the wayside. Instead, we are coming to see that the history of Earth, of all living beings, and of human civilizations is a series of stops and starts, in which equilibrium comes to an abrupt end with a sudden, severe catastrophe like the asteroid that formed the Haughton Astrobleme and the much larger object that initiated the extinction of the dinosaurs. Extraterrestrial objects are only one potential source of such changes, which include as well shifts in Earth's axis, movements of the continents, volcanic eruptions, and earthquakes. A desire to bring all this information together for the widest possible audience led me to join forces with professional science writer Robert Aquinas McNally to compose *Voices of the Rocks.*

Chapter 1 introduces the theories of uniformitarianism and catastrophism and explores the substance and implication of the paradigm shift now upon us. Chapter 2 follows one particular thread in this paradigm shift, namely my own geological research into the date of origin of the Great Sphinx of Giza. Chapter 3 pursues further the question opened by the redating of the

Sphinx: Did human civilization begin sooner than we think? The fascinating possibilities suggested by popular tales of Atlantis and other lost cities and by my own study of the Yonaguni Monument in the East China Sea form the focus of chapter 4. Pole shifts, tectonic movements, and other Earth-originated catastrophes are explored in chapter 5, while collisions with extraterrestrial objects are detailed in chapter 6. Finally, following the wisdom that what is past is prologue, chapter 7 looks at new ways of planning the human and planetary future, based on an equally new understanding of Earth's past, of human history, and of the links between the living and nonliving worlds.

Entering the twenty-first century, we face global warming, overpopulation, and chemical pollution, as well as potential natural catastrophes like earthquakes, volcanic eruptions, and collisions with near-Earth objects. Knowing what happened in the past can prepare us for the future—and increase our chances of surviving sudden environmental events. Only by separating fiction from fact and understanding truth can we face our future with clarity and shape it in wisdom. Such is the task of *Voices of the Rocks.*

1

The Changing of the Paradigm

SCIENCE ISN'T WHAT YOUR HIGH-SCHOOL CHEMISTRY TEACHER told you it is. Neither is it the old movie cliché of men in white coats, disheveled Einsteinian hair, and wild eyes, in a laboratory full of bubbling retorts and flashing electric coils, performing earthshaking experiments that reveal some long-hidden reality; nor is it a matter of blindly voyaging into uncharted realms, like Captain Kirk and the starship *Enterprise,* to discover something no one has ever seen before. The truth is a good deal more subtle and interesting—and it is critical to understanding the workings of our Earth, the significance of the changes in scientific thinking we are now undergoing, and the implications of that change for our future.

What Science Is, and How It Really Works

Science is no monolith. There isn't a single science from which all the various disciplines—e.g., biochemistry, physics, astronomy, and zoology—derive. I have friends, for example, who are devoting their research careers to the laboratory task of characterizing groups of related proteins. This empirical, highly specialized pursuit is definitely science, but it's not anything like what I do as a geologist and paleontologist who is a generalist by choice. I find myself closer to the nineteenth-century naturalists like Charles Darwin and Alexander von Humboldt. These thinkers were drawn to collecting and analyzing facts and making sense of them in a way that, by allowing us to comprehend, experience, and appreciate the order inherent in nature, provides an understanding at once intellectually useful and esthetically satisfying. My work is both empirical and philosophical.

Still, despite the obvious differences between Charles Darwin, myself, and my protein chemist friends, all scientists agree on certain points. It is these agreements that make science an enterprise as distinct and definite as writing poetry, designing a skyscraper, or deriving a mathematical proof.

All scientists share a fundamental agreement on the primacy of natural law. Fundamentally, everything we observe in the natural world depends on relationships between matter and energy governed by the fundamental physical-chemical forces and constants, such as gravity, relativity, and thermodynamics, that make up natural law. Science doesn't allow for divine intervention or miracles as explanations for natural phenomena. This doesn't mean that scientists cannot be spiritually inclined or religious—in fact, many are—but divine intervention lies outside the bounds of scientific analysis. Science requires its practitioners to be rigorous, consistent, and logical in a natural world without supernatural apparitions or divine interference.

Likewise, all branches of science share a commitment to testing their ideas against the real world. A proposed scientific theory may boast a delightful elegance, but if it does not stand up when tested against reality, then it has no value as science. To me, this is the most important aspect of science: theoretical explanations of natural phenomena that can be tested against the real world. Without such testable explanations, we don't really know which facts to look for, yet it is the facts themselves, whether derived from laboratory experiment or from observations in the natural world, that ultimately determine the worth and validity of the ideas. As we proceed more deeply into this exploration of time, catastrophe, and history, we shall see again and again how fact and explanation are interwoven and in some ways mutually dependent.

A further aspect of scientific thinking, particularly near and dear to my heart as a working scientist, is parsimony, also known as Occam's razor or the principle of economy. William of Occam (also spelled Ockham), a fourteenth-century scholastic philosopher, wrote, "Never is multiplicity to be postulated without necessity," and "It is vain to do with more what can be done with fewer." Occam's devotion to simplicity still holds. There's no reason for a scientific explanation to be more complicated than it has to be. If more than one solution or explanation is posed to resolve a given problem, the simplest explanation is the best.

Parsimony does not mean that an idea must adhere to scientific status quo. The history of science is full of examples of new, parsimonious explanations that ran counter to the scientific orthodoxy of the day. The new look at the Earth explored here is appealing in part because it offers a superior, simpler explanation than the formerly accepted view.

Science stands out, too, for being progressive. We can't say that poetry or painting gets better over time. There is no clear arc of improvement reaching from Dante to William Butler Yeats or from Michelangelo to Andy Warhol. You or I may prefer one poet or painter to another, but preference isn't the same thing as progress. Science, unlike art, does progress. We know more now than we did five years ago, and our knowledge is vastly larger today than it was five centuries in the past.

This is not to say, however, that knowledge in general can't backslide or even be lost. During the European Dark Ages of the early medieval period, practically the entire body of learning and literature from classical Greece and Rome was lost to the Western world and reintroduced through Arab scholars centuries later.

Likewise, it appears that important knowledge from ancient civilizations is only now being rediscovered, a point we will develop in detail later. Still, it remains true that over the past five hundred years science has built on itself. It has constantly changed as new ideas have replaced old, outmoded notions and we have moved toward better, more profound explanations of nature.

Thinking in Paradigms

According to the conventional view, the one popular in high-school chemistry classes, a scientific idea holds sway until a large number of contradictory or unexplainable facts—commonly called anomalies—arise to refute the old theory and support a new, competing, and superior explanation. This model is orderly, intellectual, and polite. It's also wrong.

The actual process by which science progresses is best explained in a remarkable book, *The Structure of Scientific Revolutions,* by the historian Thomas Kuhn. Kuhn distinguishes between two kinds of science: normal and revolutionary. Normal science takes place within a community of scientists who share a paradigm. A paradigm is a consensus of belief that certain solutions resolve the central problems of a particular field, be it biochemistry or astronomy. Scientists hold to the paradigm not because they rigorously examine and test it, but because they are committed to it, usually through education, professional values, and agreement among powerful institutions like foundations, universities, government agencies, and research facilities. The paradigm is right because everybody says it is right, and because it explains the facts everyone agrees are the right facts to explain. This assumption is so fundamental and unquestioned that everything inside the prevailing paradigm has a way of seeming natural and normal, while everything outside it—particularly the anomalies—seem inconsequential or irrelevant. Normal science is just what its name implies. It looks as "normal" as sunrise in the east.

Revolutionary science, by contrast, is the work of researchers outside the paradigm, those who do not accept the prevailing wisdom and focus instead on the anomalies the normal paradigm cannot explain. The revolutionaries offer up a new, competing paradigm, one that makes the previous anomalies the mainstay of their new scientific worldview. What was inconsequential and irrelevant becomes central and significant.

Since normal and revolutionary paradigms do not agree on what is central and significant, the new paradigm cannot refute the old, nor vice versa. Typically, the new paradigm succeeds if it can explain more phenomena. Although competing paradigms may explain different phenomena, one of them will usually explain more of its own phenomena than the other—and that makes it more appealing to scientists. Since the explanation of more phenomena makes for more scientific work, scientists tend to accept paradigms that give them more to do as scientists. And, given the social importance and economics of science, paradigms with social benefits are also likely to receive increasing support over time.

Kuhn's profoundly important analysis makes two points that are significant in understanding how our conception of the role of global catastrophe in Earth's history is changing. First, science makes progress by revolution rather than by refutation. An overarching scientific idea retains its position until the adherents of a competing idea push it aside, something like a band of rebels demolishing an established government and putting a new set of institutions in its place. Second, science moves not in a splendid isolation propelled by the energy of its own findings and ideas, but in relation to the large sweep of history. New scientific values arise because of changes in politics, economics, religion, and other aspects of the human context in which science is practiced. As the world changes, so do scientific paradigms.

When we set the current competition between a steady-state Earth against a fits-and-starts planet into the background of scientific thinking across history, we can see clearly how the paradigms have changed over time. We can also appreciate more thoroughly why the current debate is nothing less than the triumph of revolutionary science and the emergence of a new and sobering paradigm.

Calming the Dangerous Skies

The peoples of the ancient, preclassical world didn't practice what we think of as science. Still, they were careful observers of the natural world who noted their observations in a variety of surviving texts or passed them down in oral traditions that were later recorded in such sources as the Hebrew scriptures of the Old Testament, the *Epic of Gilgamesh,* and the Homeric poems. If these ancient works are taken at face value, the skies in those long-gone days were ominous indeed, and Earth was a most perilous place, one subject to sudden celestial destruction on a massive scale. Sodom and Gomorrah were reduced to ashen rubble by a rain of fire sent, supposedly, to punish those cities for their transgressions. Sky fire was likewise the threat that Jonah carried to Nineveh, the great Mesopotamian city, in his successful attempt to convert its people to righteousness. The Greek gods and goddesses aiding the combatants at Troy displayed themselves as terrible lightning bolts and fast-traveling stars trailing sparks. And the Babylonians believed that Earth had been washed clean in a great flood announced and accompanied by terrifying events in the heavens.

In the ancient worldview, such calamitous events were not anomalies; they counted as central events of the paradigm. The heavens were the scene of violent struggles that determined the fate of Earth and humankind. The creation myths of many of the ancient civilizations featured a primal battle pitting a celestial force of evil against a celestial force for good. For the Hebrews, the fight was between Yahweh and Satan; for the Egyptians, between Horus and Seth; for the Greeks, between Zeus and Typhon; for the Babylonians, between Marduk and Tiamat; for the Syrians, between Baal and Yam; and for the Scandinavians, between Thor and Odin. The later story of Saint George slaying the Dragon is a medieval reupholstering of the same basic mythic theme.

Looking back, we moderns may appreciate the literary quality of *Gilgamesh* or the Homeric epics, but we reject the accuracy of their content. The ancients were, we seem to believe, psychological primitives who lived in quaking fear of the heavens. They were much like children who invent monsters under the bed and in the closet, and keep themselves awake at night, worrying about fangs and claws. Yet how can we be sure of our intellectual and psychological superiority? Could we, in so smug and easy a dismissal, be missing something critically important?

The British astronomers Victor Clube and Bill Napier are certain we are. In their persuasive *The Cosmic Winter*—a work that will figure importantly later in this book as well—Clube and Napier argue that the dangerous skies of the ancients accurately represented the real world of the time, when Earth's orbit carried it through a meteor stream that often wreaked sky-borne devastation on our planet. The ancient worldview of the dangerous heavens was no infantile fantasy. Rather, it constituted a paradigm based on observation of central phenomena, one that used the available religious language of gods and goddesses to explain events larger than human life and well outside our control.

The later Greek paradigm, typified by the fourth-century B.C. Athenian philosopher Aristotle, differed strikingly from the preclassical worldview. Aristotle's view of the heavens was fixed and orderly. Earth occupied the center of a series of concentric crystalline spheres to which the heavenly bodies were attached and on which they turned—first the moon, followed by Mercury and Venus, then the sun and Mars, Jupiter, and Saturn, finally the constellations of the Zodiac, and beyond them the multitude of unnamed stars. The divine aspect of the ancient paradigm persisted in the names attached to the planets, which are called to this day after principal divinities of the ancient world. In Aristotle's paradigm of the cosmos, however, divinity was orderly and predictable, without the omnipresent and arbitrary threat befalling the Babylonians, Hebrews, and preclassical Greeks.

Typically this shift is explained as a movement from older, magical thinking to a rational, observation-oriented philosophical standpoint. It is as if the ancient world woke itself up from a bad dream, took a hard look at reality, and in its surprise announced, "The sky isn't dangerous after all. How silly and childlike of us to have thought so! Isn't it about time we got real?"

Clube and Napier argue that humans hadn't gotten any smarter over the intervening centuries. Instead, the known cosmos had changed. With the passage of time, Earth's orbit moved out of the meteor stream and the devastations of old were slowly lost to common memory. What had been chaos now became orderly and peaceful. Thus the warring celestial gods of Homer's times gave way to the fixed crystalline spheres of Aristotle's precise, benign cosmology.

The Aristotelian paradigm set the tone for European science over the next millennium and a half. Although Greek thought was largely lost during the Dark Ages of the early medieval period, it returned through the scholastic philosophers, who became reacquainted with Aristotle through Arab thinkers. The greatest scholastic of all, Thomas Aquinas—my coauthor's namesake—wedded the Aristotelian system of fixed heavens with Christian theology. Now the full force and majesty of the Church backed the paradigm of a predictable, orderly Earth and sky.

Not that there weren't anomalies, most notably comets. The sudden appearance of a bright, tailed star that traveled across the vault of the heavens and in time disappeared from view made the cosmos look much less fixed than the Aristotelian-Aquinian paradigm allowed. If the sky were set once and for all, as the paradigm required, where could a comet have come from, and how had it slipped inside the realm of crystalline spheres? One explanation was that comets were phenomena of the atmosphere, like clouds or lightning, and did not belong to the heavens proper. Yet even when a large comet appeared at the end of the sixteenth century and the great Danish astronomer Tycho Brahe showed that this body passed far beyond the moon, well outside Earth's atmosphere, many important scientists continued to cling to the Aristotelian paradigm despite Brahe's evidence. Johannes Kepler, one of the giants of astronomy, accepted that comets lay beyond the atmosphere, but he concluded that they moved through the solar system in straight lines and passed out again—strange interlopers, to be sure, but not ones that upset the accepted order of the sky. Even Galileo, that champion of intellectual honesty who was placed under house arrest by the Inquisition for championing Copernicus's idea that the sun, not the Earth, occupied the center of the solar system, dismissed comets as mere optical illusions. To him, they were no more real than mirages shimmering watery and wet in a bone-dry desert.

Isaac Newton developed his own updated version of the Aristotelian-Aquinian model. He saw the solar system as the work of a divine Creator who set the whole system in motion like a great, unending clockwork and then let it run on its own. With their unpredictable appearances and paths, comets didn't fit well in such a precise and eternal universe, which turned like so many gear wheels in the same perpetual motion. Newton steered around the issue as best he could: he hinted that God might intervene to cast a comet at Earth as an act of divine providence, but otherwise a collision between our planet and a comet simply lay outside Newton's system.

Not until well into the nineteenth century was it established that space rocks had indeed struck Earth, and that a band of asteroids orbited the sun between the planets Mars and Jupiter. As it turned out, these discoveries somewhat reassured the prevailing paradigm. Meteorites proved to be small objects, nasty enough if they crashed into your house or barn, but lacking the size to do more than very local damage. As for the asteroids, they appeared to be too far away to pose any danger to Earth. The further discovery that the sun melted many comets into harmless clouds of dust added further to the paradigmatic sense that the Earth was isolated from the heavens, safe and secure in its fixity, essentially unchanged and unchanging. It was an idea that persisted well into the middle of the nineteenth century, when Charles Darwin first presented his theory of evolution by natural selection.

Some Things Change, Some Things Stay the Same

Yet another conventional story passed down by a great many high-school—and even college—science teachers is that Darwin and the forces of scientific reason went to battle against armies of biblically inspired fundamentalists, who believed in catastrophes as signs of God's vengeful punishment and lived in holy dread of the fire next time. Darwin and his friends argued, the story continues, that only long, slow processes of evolution by natural selection could explain the appearance of new species and the fossil record of past organisms long extinct. The conventional wisdom labels this paradigm "uniformitarianism." The other side supposedly held that terrible catastrophes of the sort described in the Old Testament, like the great flood of Genesis and the plagues of Exodus, accounted for extinctions and that all species were created—and wiped out—only by the action of a God who repeatedly intervened in human and biological affairs. This point of view, with or without divine intervention, is called "catastrophism." The truth of the matter, however, is a good deal more complicated, and very much to the point of understanding the current paradigm shift.

Well before Darwin and his contemporaries, observers of nature were contemplating Earth as well as gazing at the stars and wondering why the planet had the shape that it did—plains and flats here, for example, tall mountains there. In Christian medieval times, many of the features of the planet's surface were said to result from the flood that Noah escaped in his ark filled with paired animals. Not everyone was quite sure of this notion, including the quintessential Renaissance man Leonardo da Vinci. Reflecting on the presence of fossil shells high in the mountains, Leonardo wondered how they had found their way to such altitudes. The flood, he was sure, couldn't have done it. "Things which are heavier than water do not float high in the water," he wrote, "and the aforesaid things could not be at such heights unless they had been carried there floating on the waves, and that is impossible. . . ." Leonardo suggested that the fossils might have formed when earth or mud covered shelled organisms lying on a shore or sea bottom. The mud later turned to rock, and the shore or sea bottom was raised up.

In fact, Leonardo had it right, yet in his own time his explanation seemed outlandish. Among all its many catastrophes, the Bible contained no description of such uplifting. Besides, it would take a great deal of time for mud to petrify, and everyone knew Earth had been around for far too little time for such events to have occurred. In addition, Leonardo's explanation meant that the fossil-bearing rocks had to be younger than the rocks beneath them. According to a biblical understanding of planetary history, that simply could not be.

Until the seventeenth and eighteenth centuries, most scholars and philosophers believed the Earth to be only a few thousand years old. Ancient historical records other than the Hebrew scripture went back to about 2000 B.C. Adding on the chronologies in the first five books of the Old Testament appeared to extend the time to about 4000 B.C., when God was said to have made Earth in one day of the week of Creation. Thus, all rocks had to be of the same age because they arose at the same time. Leonardo's explanation didn't fit the paradigm.

Help was on the way. In the seventeenth century, Nicolaus Steno (also known as Niels Stenson, Nicoli Stenonis, and Nils Steensen) founded the branch of geology known as stratigraphy, which concerns itself with the arrangement of rock strata. Steno recognized that all rocks are not of the same age, and that a higher stratum can be considered younger than the lower stratum on which it rests.

In the following century, James Hutton, considered the father of modern geology in English-speaking countries, took time and stratigraphy further. Trained as a physician, Hutton never practiced, living instead as a gentleman farmer in Scotland, a pursuit that taught him a great deal about soils and erosion. By examining the weathering rates of rocks and the lack of any significant changes in the Scottish coastline since the time of the Romans, Hutton concluded that time was infinitely long, much longer than the approximately 6,000 years of biblical literalist interpretation.

Hutton developed, as well, a model of Earth as a great recycling machine. Solid rock, he wrote, weathers into soil, which is then eroded by running water and carried into the oceans, where it settles as sediments that consolidate over the eons into rocks. In time the rocks are lifted above sea level, where they are again exposed to rain, and the cycle begins again. Hutton's model of an ever-recycling Earth accounted for Leonardo's mountaintop fossils. It also created a kind of eternal sameness. The cycles of erosion, sedimentation, and uplift could be repeated again and again, but the Earth itself remained more or less the same even as its materials cycled from rock to soil to sediment to rock again.

Hutton carried sameness over into his methods of observing natural processes, and for this reason he is often credited with being the first uniformitarian—a term he himself never used. Although, like Newton, a firm believer in a divine Creator, Hutton rejected godly intervention or miracles as a cause of geological phenomena: God had created Earth with an eye to giving the human race a fit place to live, but since the original Creation, He had let the planet go its own way. In rejecting supernatural explanations, Hutton argued that the key to understanding the processes of the past is observing the processes of the present, for they are one and the same. The world now is the world then, and a uniformity of conditions characterizes both what is and what was. Hutton took his uniformitarianism to the extreme of believing that the same species of organisms existed somewhere on Earth throughout geological time. Hutton was certain that even in his own day saber-toothed tigers and woolly mammoths were still to be found, if only one knew where to look.

In France, meanwhile, Georges Cuvier took a view both different and the same. He was similar to Hutton in that he separated religion from science and relied less on received dogma than on his own observations, specifically of the geology of the Paris Basin. Those observations, though, gave him a different view of Earth than Hutton's slow recycling. The Parisian rocks, in which the fossils of freshwater organisms alternated with those of saltwater creatures, told Cuvier that the sea had advanced and retreated several times. Each such movement of oceanic water must have been accompanied by a cataclysm, since the fossil plants and animals before every sea-change differed from those after. Practically everything was wiped out, then new species emerged. And this process wasn't necessarily as slow as Hutton's steady erosion and uplift. Mammoths and other large vanished mammals discovered in a frozen, well-preserved state told Cuvier that death had come upon them suddenly. Cuvier was laying the foundation of a secular catastrophism, one that had nothing to do with biblical deluges and other divine interventions.

Léonce Élie de Beaumont, a pupil of Cuvier, put forward a model to account for the evident catastrophes Cuvier described. If Earth had begun in a molten state and gradually cooled, the shrinking of the planet would cause sudden, large-scale shifts in the crust that made themselves known as earthquakes, volcanic eruptions, extensive floods, and the uplift of mountains. Given the scale and speed of these crustal shifts and the cataclysms they launched, extinctions were certain. Still, as our planet cooled over time, the periodic disruptions would lessen in intensity, leaving contemporary humans more secure than the vanished species that came before—in some manner, still guaranteeing the security of the original Aristotelian-Aquinian paradigm, at least for those of us lucky enough to live in the present.

The Missing Links

Beaumont's model was widely accepted among scholars and naturalists on the Continent and in Britain in the early nineteenth century. Yet his catastrophist idea was soon pushed aside by a thoroughly uniformitarian paradigm that dismissed cataclysm as a mechanism of any importance at all. Fascinatingly, scientists of different persuasions looked at the same body of facts and came away with utterly different, mutually exclusive explanations.

Both uniformitarians and catastrophists did substantial geological fieldwork, and the data they gathered laid the basis for constructing a history of Earth. The geological record was divided into periods that were grouped into eras, with a change in the fossil record indicating the boundary dividing one era or period from the next. One of the most dramatic of these boundaries came at the end of the Permian period. Another marked the end of the Cretaceous period, and is now associated with the extinction of the dinosaurs—animals then little known to science. Such obvious evidence made it clear to all geological fieldworkers that abrupt shifts in the fossil records of plant and animal communities existed. The issue was what to make of them.

Catastrophists, like Cuvier and Beaumont, considered the sharp fossil breaks to be evidence of cataclysmic changes in Earth itself, not the workings of a vengeful and interfering God. To the catastrophists, such sharp changes were the central facts scientists needed to explain. The uniformitarians didn't agree.

The most articulate and persuasive of them was Charles Lyell, who expanded on Hutton's work. Originally a trial lawyer, Lyell used all of an attorney's persuasive skills to win the scientific world to his point of view.

Lyell believed deeply in a steady-state Earth, despite the abundant geological evidence pointing to sudden fits and starts. And he believed too, as did Hutton, that the history of Earth's surface could be explained in terms of the processes currently operating and observable. Since earthquakes, tidal waves, volcanic eruptions, and other cataclysmic events in the contemporary era were all merely local events, Lyell held that there was no reason to explain events by means of great cataclysms no one had ever witnessed or recorded.

But what of the dramatic evidence offered by the Permian and Cretaceous extinctions? Ah, Lyell argued, this isn't what you think it is; explaining extinctions as evidence of catastrophe is overly simplistic. Were you more open to right thinking, you would see that the real problem is an absence of evidence for the gradual transition we all know must have occurred. What with erosion and other gradual processes of decay, the clues have weathered away. Sudden shifts are illusions. Gradual change, with certain links of evidence missing in the chain of transformation to give a false impression of abruptness, is obviously the truth of the matter.

Remarkably, Lyell was a staunch creationist, a devout Christian who believed God had made Earth and all its beings, and who dismissed the evolutionary ideas then emerging in European scientific circles. Remarkably, too, his uniformitarian paradigm triumphed over secular catastrophism, largely because he simply waved his philosophical wand and made the telling geological evidence of mass extinctions irrelevant.

Charles Darwin incorporated certain of Lyell's ideas into his classic *On the Origin of Species.* Lyell was uncomfortable with Darwin's evolutionary thinking, and the once-cordial relationship between the two men cooled when Lyell realized that Darwin considered humans as much the product of natural selection as any other species—an idea that the more traditionally religious Lyell would not accept. Yet Darwin, like Lyell, believed strongly that rapid change was an illusion and that the change of one species into another was glacially slow, the result not of some sudden cataclysm, but of the gradual accumulation of many small mutations and selections. In the final chapter of the *Origin,* Darwin wrote, "As natural selection acts solely by accumulating slight, successive favourable variations, it can produce no great or sudden modification; it can act only by very short and slow steps." Extinctions did occur; the evidence of species passing away and new ones arising was undeniable. Still, it took a long, patient while for such events to happen, all of them explainable in terms of natural selection's apparent slowness. "The extinction of species and of whole groups of species, which has played so conspicuous a part in the history of the organic world, almost inevitably follows on the principle of natural selection; for old forms will be supplanted by new and improved forms," Darwin wrote.

Certain that the process was slow and gradual rather than abrupt and sudden, Darwin revisited Lyell's argument about absent clues. He noted that fossils can be destroyed by a great many causes and concluded that "All these causes [for the destruction of fossils] taken conjointly must have tended to make the geological record extremely imperfect, and will to a large extent explain why we do not find interminable varieties, connecting together all the extinct and existing forms of life by the finest graduated steps."

In the century and a half since Darwin's formulation of the theory of evolution by natural selection, his ideas have been revised and updated, yet key facets of the paradigm have remained set. Lyell's steady-state Earth slowly disappeared from the scene, as geologists became increasingly aware of the reality of ice ages and continental drift as global events. Still, the shifts associated with these phenomena were thought to be extremely slow, fitting therefore with the overarching gradualism of Darwin's natural selection theory.

Under pressure from the success of the uniformitarian paradigm, the scientific catastrophism of Cuvier and Beaumont faded away. Only a biblical catastrophism with a distinctly medieval flavor remained. In Newton's time, for example, some scholars saw comet collisions as possible sources for Greek myths like Phaethon's ride in the fiery chariot, as well as for biblical accounts of the flood of Noah, the plagues that persuaded Egypt to let the Israelites go, and similar events. Churchmen and philosophers of nature wondered in print whether such events didn't in fact prove the truth of biblical prophecies, and showed, too, that God worked through the skies to render His judgments on sinful humans. This kind of thinking reappeared in the work of biblical catastrophists in the nineteenth century. One of these was the German linguist Johann Gottlieb Radlof, who argued that the Earth had been struck by a comet fragment called Phaethon, and that another piece of the same comet had become the planet Venus. The most successful and notorious of the biblical catastrophists was Immanuel Velikovsky, who resurrected some of Radlof's ideas in the twentieth century. A psychoanalyst by profession, Velikovsky argued, in his 1950 book *Worlds in Collision,* that Earth had suffered an axis shift as a result of a close encounter with a comet, and that the planet Venus had hurtled past Earth in 1500 B.C.

Utterly lacking physical evidence for the catastrophes he described, and basing his work on a selective analysis of end-of-the-world mythologies, Velikovsky drew fierce attack from academic scientists. In part, they were bothered by the lack of any data, other than a literalist reading of ancient mythologies, to support his position. Yet the fierceness of the response showed how ingrained the uniformitarian paradigm was, and how unwelcome dissenting views had become. Since the time of Aristotle and his scholastic descendant Aquinas, Western thinkers had largely held to the fundamental paradigm of an Earth free of catastrophes, splendidly isolated from the heavens, and subject only to the slowest of changes. It would take a good deal more than Velikovsky to shake this paradigm's long-standing grip on the scientific mind.

Dead Dinosaurs, the Big Bang at Chicxulub, and Catastrophism's Resurrection

To give the scientific establishment of Velikovsky's time its due, the contemporary database lacked strong evidence of any celestial or terrestrial process destructive enough to massively disrupt our planet. Comets were considered gaseous rather than solid, and therefore no potential danger to Earth until 1951, when F. L. Whipple proposed the "dirty snowball" theory of cometary composition. Likewise, until the 1950s most astronomers were confident that only a few meteors* of a size capable of doing widespread damage had ever survived their high-speed trip through the atmosphere and actually struck Earth. Although large, craterlike structures had been described, no meteorites of more than middling dimensions were discovered at these sites. The issue, it turned out, wasn't an absence of impacts, but the vaporizing explosion that largely destroyed the space rock. Still, it is a long intellectual leap from the theoretical possibility of Earth running into a big leftover from a comet or a wandering asteroid and an overturning of the uniformitarian paradigm for a new catastrophist view.

The major anomaly in the uniformitarian model was the mass extinctions, particularly the Permian and the Cretaceous. In 1962, Norman Newell gave an address as the retiring president of the American Paleontological Association in which he argued that the extinctions weren't an illusion produced by missing evidence in the geological record—a key proposition of Lyell and Darwin's gradualism. Nor, Newell said, was it simply a matter of new species outcompeting former forms, something like new and

---

* **Technically, a *meteoroid* is any particle of matter in space, ranging from bits of dust to asteroids. A meteoroid that strikes Earth's atmosphere, becoming incandescent in the process, produces the phenomenon known as a *meteor*, or shooting star. In common parlance, however, the term *meteor* is generally used to refer to both the observed phenomena of shooting stars and to the solid meteoroid objects. A *meteorite* is the portion of a meteor or meteoroid that reaches Earth and collides with the surface.**

improved models shouldering aside old fogies. If that were the case, fossils of old and new species should be found mixed together. Rather, the former groups of organisms completely disappeared, and a significant delay ensued before new species arose to occupy the vacated spaces. Still, Newell thought the process was quite slow, taking on the order of 6 million years following each mass extinction.

A few other scientists developed complementary hypotheses that extinctions were sudden rather than slow, and possibly the result of collisions between Earth and large asteroids or comets. Though these ideas escaped the vituperative response that met Velikovsky, they were still ignored. Good solid evidence to support such ideas didn't exist, and they seemed too outlandish to pursue.

Then University of California geologist Walter Alvarez asked his father, the Nobel Prize–winning physicist Luis Alvarez, to join him in investigating the Cretaceous extinction, which occurred some 65 million years ago. This was an enormous event that wiped out one in four of all animal families then extant, including the dinosaurs, which had flourished for the prior 150 million years. All the winged and large marine reptiles also disappeared, seemingly in short order, as did the ammonites, a large group of marine cephalopods related to today's octopi and squids. In many geological sections, the boundary between the strata of the Cretaceous is marked off from the subsequent strata of the Tertiary by a thin layer of clay less than half an inch thick, the so-called K-T boundary. Interested in any information about the process of extinction that could be found in the K-T boundary, the younger Alvarez wanted his physicist father to help him study it in Gubbio, Italy, where the clay layer is easily accessible.

The Alvarezes' investigations revealed that the K-T boundary clay in Gubbio was rich in iridium, a metal almost completely absent on the Earth's surface, but found in greater abundance deeper within the planet. Subsequent studies of the K-T boundary in Denmark and New Zealand uncovered the same high concentration of iridium, ruling out the suggestion that the iridium was unique to Gubbio. This meant there were only two possible sources for the metal: either volcanic activity, which would bring the element in molten form from deep inside Earth; or an asteroid, which, since it would date from the same period as Earth's interior, would be similarly rich in iridium. None of the K-T sites showed evidence of volcanoes, leaving the asteroid model as the only workable alternative. In 1980 the Alvarezes published a paper hypothesizing that an asteroid on the order of six or seven miles wide had slammed into Earth, resulting in a massive explosion and catastrophic climate change that wiped out the dinosaurs.

Despite the extraordinary credentials of the paper's authors, the vast majority of scientists were at best skeptical. Beverley Halstead, a British paleontologist, sought to tar the Alvarezes with the same brush as Velikovsky, dismissing their idea as theologically inspired: "The asteroid or giant meteor explanation has the great popular appeal of high drama.... Such theories are clearly an advance on invoking the wrath of a Deity but not very much."

Further fieldwork that identified high iridium in K-T clays from over fifty sites showed the Alvarezes were onto something. Many of the K-T clays also contained materials, such as shocked quartz grains, that are associated with meteor impacts.

A competing explanation for these data was offered: massive volcano activity. During a time period that roughly matches the K-T boundary, some 200,000 square miles in what is now west-central India—an area the size of Washington and Oregon combined—was covered with lava flows 6,500 feet deep. The total amount of volcanic material in these flows, which are called the Deccan Traps, is estimated to be more than a half-million times the volume of material released by the Mount St. Helens eruption. Explosive volcanic activity on a large scale has major environmental effects, such as acid rain, reduced alkalinity of the oceans, changes in the atmosphere around the globe, and depletion of the ozone layer, all of which would have profoundly affected many kinds of living beings, both plant and animal.

Still, explaining the extinction and the puzzling features of the K-T boundary clay as a result of the Deccan Traps vulcanism didn't quite fit. The lavas of the Deccan Traps do contain iridium, although at far too tiny a concentration to explain the striking presence of this element in the K-T clay. And there was a timing problem. The K-T clay represents a period of from only one to several years. The Deccan Trap lava flows began several million years before the K-T clay and continued for several million years after. Curiously, however, recent research shows that 70 percent of the flow was laid down over a period of less than a million years, straddling the narrow time of the K-T clay. Could there be a connection? Did a major impact cause or accelerate the Deccan Traps?

Part of the answer came in 1990, when a large impact crater of precisely the correct age was identified at Chicxulub in what is now Mexico's Yucatán Peninsula. The asteroid or comet, which was several miles across and about the size of Mount Everest—at least ten times as big as the space rock that produced the Haughton Astrobleme—slammed down in shallow water off the ancient coastline. The resulting explosion was unimaginably massive, on the order of 10,000 times greater than the simultaneous detonations of all the world's current stockpile of nuclear weaponry. The effect was equally devastating. The impact explosion gouged out a crater over 110 miles wide and 20 miles deep, killing everything for thousands of miles around. Earth rocked with powerful earthquakes, and coastlines were inundated by towering tsunamis. The huge volumes of dust raised into the upper atmosphere by the explosion blocked sunlight for months, years, perhaps decades, slowly killing off many of the land plants and the animals that fed on them and starving the predators in turn. Over time and the turning of the eons, the Chicxulub crater filled with sediment and eventually became part of Yucatán.

The Chicxulub impact may have been only one of a cluster of such collisions that occurred within a short period, an idea we will explore further in chapter 6. By penetrating into the mantle layer of Earth, the impacts could have contributed to the volcanic processes that formed the Deccan Traps, and the iridium found in the Indian lavas may also have arisen from these collisions. It is also possible that by unfortunate coincidence an asteroid hit Yucatán at just the time when extensive volcanic activity was taking place in India, and that the two processes, the one sudden and the other gradual, together created complementary cataclysmic conditions causing the mass extinction. Clearly, the interaction of the Chicxulub impact with the extensive vulcanism of the Deccan Traps still needs to be sorted out.

Yet one thing is sure: Only a catastrophist paradigm accounts for the evidence. The discovery of the Chicxulub crater and its association with the K-T boundary has upended the uniformitarian paradigm and changed scientific thinking. A new catastrophism is becoming an established fact of life and Earth science.

Evolution and the Catastrophe Question

When we compare the catastrophist paradigm with the scientific criteria discussed at the beginning of this chapter, we can see how well this new model advances our thinking. To begin with, catastrophism fits natural law. There's nothing divine about a comet hitting Earth or a massive outburst of volcanic activity. The geological record as well as astronomical observations, such as the collision of the comet P/Shoemaker-Levy 9 with Jupiter (which will be discussed at length in chapter 6), support these phenomena as ordinary and expectable. Likewise, catastrophism is more parsimonious than uniformitarianism. Lyell's and Darwin's argument about missing links in the geological evidence is much more fanciful than accepting the abruptness of boundaries on its face. And catastrophism is progressive; it adds to our comprehension. Catastrophism gives us a superior handle for understanding more of the phenomena of the natural world. Evolution by natural selection serves as a good example.

As we have seen, Darwin's original model proposed a slow, gradual accumulation of small changes that over time molded a new species from an old. But as we have also seen, this hypothesis didn't fit very well with the fossil record, in which some species suddenly vanished and, after a delay, new forms appeared with equal suddenness. Even before the details of the Chicxulub impact became clear, paleontologists like Niles Eldredge and Stephen Jay Gould were proposing a new model of evolution that moved not slowly and gradually but in fits and starts. This new theory is called "punctuated equilibrium," a term that conveys an image of steady states regularly interrupted by periods of rapid, even radical, change—something like long sentences broken by commas, semicolons, and the occasional abrupt period.

Catastrophism dovetails beautifully with punctuated equilibrium. In a steady-state world, all or nearly all of the ecological niches are taken up by existing plants and animals. If new species were to arise, the world would have little room to accommodate them. In such a state, innovation is unusual; things and beings tend to stay pretty much the same. Then there comes a period of severe environmental stress—caused perhaps by an asteroid impact, a change in planetary temperature, or massive volcanic activity—that kills off a large proportion of the existing organisms. Smaller gene pools are more influenced by genetic change than larger ones, so that change accelerates in the organisms that do survive. And, with many ecological niches vacated by the cataclysm, these surviving, fast-changing organisms have new spaces to move into—novel food sources to exploit, unoccupied habitats to colonize. Obviously, creatures that are genetically nimble and able to adapt quickly are the most likely to succeed under the new conditions created by the catastrophe.

The model of punctuated equilibrium requires periodic catastrophes. The sources of those catastrophes make little overall difference; only their destructiveness counts. Without some major upset in the system, the steady state would continue steadily on, changing only very slowly, if at all. The result would be an inevitable stagnation in evolutionary change. If that massive asteroid hadn't slammed into Chicxulub 65 million years ago, dinosaurs might still be the largest land animals, mammals would be restricted to a few small rodents and insectivores that dared to forage only at night, and we humans would never have arisen. Catastrophe primes the evolutionary pump; it provides the energy and the opportunity that moves natural selection in new directions—probably including, as it turns out, our own.

By showing how catastrophes prove "useful" to evolution, catastrophism helps explain a great deal of paleontological information that used to seem unexplainable. Yet in this understanding there is a profoundly discomfiting note. We live now in a world profoundly different from the one inhabited by the last of the dinosaurs, 65 million years ago, yet the physical and chemical constants that make up natural law remain the same. The processes of the past and the present are one and the same. What befell the dinosaurs could just as easily have befallen us, not only in the far distant eons that set the stage for our evolution as a species, but also in more recent times. Clearly, such a catastrophe is theoretically possible, even expected—but did it happen, when, and what effect might it have had on our forebears?

2

A Shape with Lion Body and the Head of a Man

WHEN WILLIAM BUTLER YEATS CONJURED THE IMAGE OF THE Great Sphinx of Giza in his magnificent poem "The Second Coming," describing it as "A shape with lion body and the head of a man / A gaze blank and pitiless as the sun," he was harking back to this mythical creature's ominous import in classical antiquity. In that long-ago world, the sphinx was linked inextricably to unanswerable questions, the unstoppable passage of time, and the threat of destruction.

The most telling example is the Greek legend of Oedipus. A particularly monstrous sphinx beset the city of Thebes, devouring every passerby who could not solve its riddle: "What creature goes on four feet in the morning, on two at midday, on three in the evening?" Oedipus recognized the answer: "A human being. In childhood, one crawls; in adulthood, one walks boldly; in old age, one hobbles with cane or staff." That answer lifted the curse, and Thebes, in gratitude to the unknown wanderer, made Oedipus king and married him to the prior king's widow. In this tale, however, no one lived happily ever after, as Oedipus discovered that he had unknowingly killed his own father and married his mother. In the end, all were destroyed—Oedipus, his mother and wife, and all his children—and the city of Thebes devolved into a brutal civil war.

I, too, came to the Great Sphinx to answer a question, not as a poet or an aspiring king, but as a scientist probing an issue of time. What I found there opened my eyes to a new avenue of scientific thought about human origins and the role of catastrophe.

Asking the Unaskable

Even before I met John Anthony West, my own interest in ancient history had taught me something about the Great Sphinx of Giza. This immense statue—66 feet high and 240 feet long—rests in the shadow of the Great Pyramid, the only survivor among the Seven Wonders of the ancient world. Carved from limestone bedrock to resemble what Yeats described, the lion-bodied, human-headed monument sits on the edge of Egypt's Giza Plateau, above the Nile River a few miles west of Cairo on the edge of the Sahara, within the same complex of ancient structures as the three classic pyramids of the Old Kingdom's Fourth Dynasty. Since the 1930s, most Egyptologists have dated the construction of the Sphinx to the reign of the Pharaoh Khafre (also known as Chephren), in approximately 2500 B.C.

John Anthony West wasn't so sure about that. Often described as an "independent" Egyptologist—which is to say a man without formal academic training or credentials in the field—West has long been fascinated with ancient Egypt, making his living by conducting tours to the country and writing popular books on the subject. West came to me in 1989 through a mutual friend, Robert Eddy, a former professor of rhetoric at Boston University, because he wanted the help of a geologist in solving a puzzle about the Sphinx.

In the course of his research some years earlier, West had come across the work of René Aor Schwaller de Lubicz. An Alsatian philosopher and mathematician, Schwaller de Lubicz conducted an extensive survey of the Temple of Luxor in Egypt between

1937 and 1952, and uncovered geometrical relationships within the structure that no one else had detected. He concluded that much more was going on in ancient Egypt than most conventional scholars were willing to admit, and he argued that its civilization boasted a sophistication we moderns do not understand. Although Schwaller de Lubicz's ideas were reviewed and supported by the orthodox Egyptologist Alexandre Varille, the academic establishment ignored his studies as unimportant, and dismissed them as self-evidently wrong.

West, though, found much of value in Schwaller de Lubicz. Among other things, he was intrigued by the observation that the Great Sphinx showed a weathering pattern different from the other structures at Giza. Schwaller de Lubicz suggested that the difference might be due to water erosion rather than windborne sand. This possibility caught the attention of West, who had an intuitive feeling that the roots of Egyptian civilization reached much farther back than 2500 B.C. The climate in North Africa was vastly wetter in the millennia preceding the Old Kingdom's Fourth Dynasty than it was during the Fourth Dynasty itself. Might the Sphinx have been exposed to flooding during this particularly wet period? If that proved to be the case, West realized, then the Great Sphinx was constructed well before 2500 B.C.

West didn't have the scientific training or tools to test his idea with the rigor it deserved. He needed a geologist. That was how he came to me.

As I listened to West outline his notion of the Sphinx's age, I was skeptical. After all, the evidence linking the Great Sphinx to the reign of Khafre was circumstantial but consistent. Of the three great pyramids, the one attributed to Khafre sat closest to the Sphinx. Statues of Khafre found in the nearby building called the Valley Temple added to the association. So did a later, New Kingdom inscription on a granite pillar, or stela, positioned between the paws of the Sphinx, whose damaged hieroglyphics contained the first syllable of Khafre's name. Further, some scholars said they saw the face of Khafre in the Sphinx's blank and pitiless gaze. I had no reason to believe the geological evidence would do anything other than confirm the Egyptological consensus. I told West I'd take a look, and probably disprove his notion, if he could get me to Egypt.

West did just that the following summer, and at eight-thirty in the morning of June 17, 1990, I was standing in front of the Sphinx. While some people describe a powerful mystical experience on encountering this monument, I was unmoved, holding to a scientist's cultivated dispassion. As I spent the following days taking careful observations, however, I developed a deep admiration for the Great Sphinx of Giza as an ancient relic, a survivor from a long-lost world strikingly different from ours, one able to speak now only in stone.

The Voices of the Ancient Rocks

I soon realized there might be something to West's intuition about the age of the Sphinx.

The statue sits not, as people often imagine, in majestic repose upon the highest point of Giza, but in a hollow on its eastern edge, so that only the head and the top of the back project above the general elevation of the plateau. This hollow, known as the Sphinx enclosure, is in actuality an ancient quarry from which the ancient Egyptians carved away the bedrock limestone to form the lion-bodied statue. The nearby Valley and Sphinx temples were built with blocks of this same limestone covered with granite facing-stones called ashlars, which had been transported down the Nile from Aswan.

The more I looked at the two temples built from the same limestone as the Great Sphinx, the more I became convinced that they were constructed in two steps, separated by considerable time. The limestone blocks cut from the Sphinx enclosure showed the uneven surface expected from long-term weathering. The granite ashlars had actually been shaped to fit these undulating surfaces. In some places, ancient Egyptian masons had attempted to even out the weathered surface, but they didn't take off enough to leave a perfectly smooth wall.

Old Kingdom inscriptions cut into the granite ashlars date these stones to the general time of Khafre—and this apparently reliable fact raised a problem. It was unlikely that the Egyptians had used weathered limestone to build the two temples. If we make the reasonable assumption that the builders began with smooth, unworn blocks, the original limestone would have had to be exposed to the elements for a long while before it revealed the pattern of wear evident under the granite. This meant that the limestone predated the granite by a considerable period, and that the temple had been built not during the Fourth Dynasty but long before. Khafre or one of the other Old Kingdom pharaohs had merely added the granite to the original, weatherworn temples at a later date, probably intending to refurbish or remodel these old structures.

Of course, since the limestone blocks in the two temples were cut in the course of carving the Sphinx, they and the Sphinx had to be of the same age. And if the limestone blocks were older than Khafre and the Old Kingdom, then so was the Sphinx.

On this first trip to Egypt, I could do no more than play world traveler and examine the temples, enclosures, pyramids, and causeways of the Giza complex from the respectful distance allowed to the many tourists who visit the site. Yet I saw enough from even such a poor vantage point to convince me that a much closer scientific investigation of the Sphinx was warranted. Specifically, I needed to get into the Sphinx enclosure and examine the statue itself.

When I returned to the United States, I drafted a lengthy proposal to the Egyptian Antiquities Organization (EAO), which then administered all such research, and requested permission to perform geological studies of the rocks forming the Sphinx and the temples. I promised to be "noninvasive"; that is, I didn't plan to take samples of the limestone. I also wanted to look at the rocks under the surface by means of seismic testing. The EAO granted permission.

In April 1991, West and I returned to Giza, bringing with us Thomas L. Dobecki, a seismology expert then with a consulting firm in Houston, who came along to help with the technical aspects of the seismic study. We gathered data by positioning a steel plate at predetermined points, then striking the plate with a sledgehammer. This energy traveled as sound waves through the surface gravel, sand, and uppermost rock layers, then reflected off whatever was underneath. Special microphones called geophones picked up the reflected sound waves and transferred the information to a sophisticated portable seismograph for storage on computer disks. Through analysis of the volumes of data we collected, we could later create a cross-section of the subsurface structures without in any way damaging the Sphinx or its surroundings.

Dobecki, West, and I discussed our preliminary findings with scientific colleagues at the University of Cairo, then we returned to the United States to study the data more carefully. In June I went back again to gather more data and verify the information already gathered.

The more I studied the results, the more they confirmed what I had suspected from the two-stage construction of the Valley and Sphinx temples. Not all the structures of the Giza Plateau were built during the Fourth Dynasty, as was generally accepted. Some of them, including the Great Sphinx, had already been around for a long, long time before Khafre mounted Egypt's throne.

The Marks of the Old Rains

One of the common methodologies of geology is analysis of the modifications made to exposed rock surfaces by weathering, erosion, and similar processes as a way of determining relative ages. This approach proved key in studying the Sphinx.

It was apparent to me that two very distinct weathering processes had affected the Giza structures. In an arid environment like that of Egypt, where the wind blows steadily during certain months of the year, wind-driven sand wears away softer rock, leaving gaps between harder layers. This kind of weathering was most prominent in the structures dated unambiguously to early and middle Old Kingdom times, the period stretching from 2600 to 2300 B.C.

The Sphinx showed some wind erosion, particularly on the head and upper back, which sit above the ground level of the plateau. However, the Sphinx also displayed obvious and extensive wear from precipitation. Rock worn away by rain has a rolling, undulating surface, often displaying distinct vertical crevices. This kind of erosion is well developed and prominent on the body of the Sphinx and within the Sphinx enclosure, where the weathering reaches from over three feet to more than six feet deep below the surface. Even though certain of the Giza structures are built from the same kind of limestone as the Sphinx, none of them show the same degree of precipitation-induced weathering.

Interesting corroborative evidence come from the Saqqara Plateau, located about ten miles from Giza. At Saqqara, a number of fragile mudbrick tombs called mastabas are dated indisputably to the First and Second dynasties, several hundred years earlier than the Sphinx's putative 2500 B.C. origin. None of these tombs bears the marks of the kind of rain-caused weathering seen on the Sphinx and the Sphinx enclosure. In fact, the mastabas were preserved by being buried in dry, windswept sand.

The Saqqara mastabas indicate that the climate has been parched and dry in northern Egypt since circa 3000 B.C. If the Sphinx had been built at this time, then, like the other Giza and Saqqara structures known to have been built during this period, it would show primarily wind erosion. The prominent precipitation-induced erosion on the Sphinx indicates that it must have been built when the climate was much wetter.

Egypt hasn't always been the desert it was in Old Kingdom times and remains today, with rainfall scarce and scant. Research into climate history shows that beginning sometime in the time span reaching from 10,000 to 8000 B.C.—the period associated with the end of the most recent ice age—Egypt was much wetter, with the contemporary desert a green savanna. This moist climate, sometimes referred as the Nabtian Pluvial, lasted until somewhere between 3000 B.C. and 2500 B.C., when the arid conditions that still prevail set in.

It seemed apparent to me that the Sphinx had been carved before the current desert climate was established and that the extensive precipitation-induced weathering was due to the heavy rains that fell long before the Old Kingdom, perhaps even before Egypt's First Dynasty in 3000 B.C. The seismic data pushed the date back even farther.

The information Thomas Dobecki and I gathered showed that the Sphinx enclosure is weathered unevenly. The north, south, and east floors of the trench surrounding the east-facing Sphinx are weathered to a depth of six to eight feet below the level of the enclosure's currently exposed surface. On the monument's western end, the Sphinx's rump, the weathering extends to only four feet. This finding surprised Dobecki and me. Since all of the limestone exposed in the trench belongs to the same stratum, we expected to find even weathering—assuming, of course, that the enclosure was quarried at the same time. Then again, maybe it wasn't. The data could be explained if the Sphinx had been carved in at least two stages, with the rump being excavated later than the remainder of the monument.

Now a likely picture emerged. When Khafre built the second pyramid in circa 2500 B.C. and added the granite ashlars to the Valley and Sphinx temples, he also finished carving the rump down to its current level, even with the floor of the enclosure, again refurbishing an already-existing monument. Originally, I suspect, the Sphinx was intended to appear like an organic part of the Giza Plateau, rising directly out of the bedrock. A similar, though much later (New Kingdom, Eighteenth Dynasty) example of this style is seen in the Temple of Hatshepsut, at Deir el-Bahri (also known as Deir el-Bahari) on the West Bank of Thebes, which is partly carved from and built into the local cliff, and esthetically integrated with its natural surroundings. In its first incarnation, the Sphinx may have had a similar appearance of blending in with its surroundings, at least when viewed from the rear. Khafre changed that, by excavating down to a level equal to the enclosure's surface depth on the other three sides.

Khafre very possibly did other repair work to the Sphinx as well. In a number of places where rain erosion had worn away the original limestone, blocks of stone were hoisted into place to fill the gaps. According to a number of Egyptologists, these replacement stones show the masonry techniques of the Old Kingdom. If the Sphinx was carved in the Old Kingdom, why did it need such extensive repair so soon thereafter? Didn't it make more sense to see the Sphinx as a much older, already-weathered structure incorporated into the Giza complex by the pharaohs of the Fourth Dynasty and repaired and refurbished by them?

A number of researchers have noticed the precise geometrical relationships between the pyramids and the Sphinx. University of Chicago Egyptologist Mark Lehner made the point that "considerable forethought went into the location of the Sphinx in relation to the rest of the Khafre complex." He is right about the geometry, but I suspect he has the time relationship and artistic intention backward. The Khafre complex was built to merge with and complement the Sphinx, not vice versa.

Still, dating the Sphinx to the period before Khafre left unanswered a major question: Just how old is it? Again, the seismic data suggest a workable answer.

It seems likely that the western end of the Sphinx enclosure was completed at the time of Khafre and has therefore been subject to less weathering than the previously exposed surfaces at the sides and front of the structure. Since this side and front weathering is 50 percent to 100 percent deeper, it is reasonable to estimate that the excavation at those points is 50 to 100 percent older than the now 4,500-year-old work at the Sphinx's rump. This line of thinking dates the original excavation of the Sphinx to somewhere on the order of 7000 to 5000 B.C., a figure that fits with the climatic history revealed in the rain erosion patterns.

As earthshaking a calculation as this has proved to be to conventional Egyptology, it is in fact a conservative estimate that represents only a minimum age. The reason is that the weathering may proceed nonlinearly. The rate of wear can slow as time passes, because the older weathered material overlies and protects the rock underneath. If this phenomenon holds true for the Giza limestone, then the Sphinx may have been carved even earlier than 7000 B.C.

Convention and Circumstance

We scientists are trained in a certain caution. Whenever we come up with findings that fly in the face of current wisdom, we check everything again and again before venturing out in public with an announcement of "revolutionary" results. I was well aware of the need to be as sure as I could be of the accuracy of my data and hypotheses. And I was aware, too, that as a geologist I was crossing academic boundaries into the barony of Egyptology.

To avoid being seen as an upstart interloper, I carefully examined the accepted arguments used to date the Sphinx to Khafre's reign in approximately 2500 B.C. I wanted to be very sure I wasn't missing something the acknowledged experts in the field took for granted.

The standard attribution of the Sphinx to 2500 B.C. can be traced to the research of Selim Hassan, who in 1949 published findings from his fieldwork during the 1930s. Interestingly, even

Hassan said that his line of reasoning and the evidence he marshaled were only circumstantial. He could not demonstrate definitively that Khafre's workers had carved the Great Sphinx.

Before Hassan, Egyptologists debated the Great Sphinx's age back and forth. E. A. Wallis Budge, well known as the translator of *The Egyptian Book of the Dead,* was of the opinion that the Sphinx was older than Khafre. And Sir Flinders Petrie, one of Egyptology's founding patriarchs, considered the structure predynastic, more ancient even than the Old Kingdom.

Hassan—who maintained that Petrie changed his mind late in his career—did correctly perceive that the Great Sphinx, the Valley Temple, the Sphinx Temple, and the Khafre pyramid belong to one overall ground plan. Therefore, he assumed that all the structures had to be built at approximately the same time. That is hardly the only explanation, however. The Khafre pyramid complex could well have been designed to fit in with a Sphinx that was already there.

Contemporary Egyptologists refer to three additional pieces of evidence in dating the Great Sphinx. The first is a strikingly beautiful statue of Khafre recovered from the Valley Temple in 1860. The presence of this statue alone, however, is hardly convincing evidence that Khafre was the builder of the temple. He could just as easily have had the statue placed in an existing temple, perhaps to appropriate its sacred energy for himself.

The second piece of evidence is the New Kingdom stela erected between the paws of the Sphinx by Pharaoh Thutmose IV (also known as Tuthmosis IV) in circa 1400 B.C., when the monument was dug out of the sand that had buried it for much of the time since the end of the Old Kingdom. First excavated in the nineteenth century, the inscription on the stela was reported to have included the first syllable of Khafre's name. Unfortunately, the portion of the inscription bearing this hieroglyph has flaked away and can no longer be studied firsthand. If the inscription did indeed mention Khafre by name as the creator of the Sphinx, it is the only such ancient inscription to do so—a curious omission among a people as given to the cultivation of royal ego as the ancient Egyptians were. And the hieroglyph in question may not have referred to Khafre anyway. The symbol for the syllable *khaf* is found in a number of Egyptian words besides the pharaoh's name.

The third and final piece of evidence is the alleged similarity between the face of the Sphinx and that of Khafre. This is a relatively recent idea, and certainly the least convincing argument of all. For one thing, the Sphinx's face has been badly damaged. Trying to determine what it once looked like is as much a matter of conjecture and artistic intuition as science. An example is Mark Lehner, who used a computer program to reconstruct the appearance of the undamaged Sphinx and felt that the face "came alive" when he gave it Khafre's features. In other words, when Lehner made the Sphinx look the way he thought it should, then it looked the way he thought it should. Such reasoning is, of course, circular.

It also has no bearing on what my geological and seismic findings revealed. From the first time I saw the Great Sphinx, and particularly after I was allowed to inspect the head firsthand and up close, I have been convinced that the Sphinx's current head isn't the original. Relatively recent tool and chisel marks, as well as the appearance of the stone itself, indicate that the current head is a recarving from an original (which may have represented an animal rather than a human). This recarving hypothesis also helps explain the head's obviously small size in relation to the body, a disruption of proportion unusual in Egyptian monuments and distinctly unlike other extant statues of sphinxes. Such a recarving would fit with the overall refurbishing and construction project attributed to Khafre. It is very possible that his workmen recarved the statue to look like him.

Or perhaps whoever reworked the head wanted it to look like someone else entirely, a person or divinity whose identity remains unknown. In October 1991, Frank Domingo, then a senior forensic officer with the New York City Police Department, traveled to Egypt to do what forensic officers do—develop an image of the Sphinx's face as if it were the suspect in a crime. Domingo concluded that the faces of Khafre and the Sphinx are different. Further, he was convinced that the faces actually represent people of different races, with Khafre's face looking more white or European and the Sphinx appearing African or Nubian. The Egyptians of the Old Kingdom are thought to have arisen from an ancestral mix of Europeans and Africans. If Domingo is right about the Africanness of the Sphinx—and I think he is—the face would indicate that even the recarved face dates back to a time well before Khafre, when Africans predominated in Egypt.

It is clear that all the evidence attributing the Great Sphinx of Giza to Khafre is circumstantial. Curiously, other circumstantial evidence points, like the African face, toward an earlier origin.

The so-called Inventory Stela (also known as the Stela of Cheops' Daughter), which dates from the seventh or sixth century B.C. and which purports to be a copy of an Old Kingdom text, states that the Great Sphinx was already in existence during the reign of Khufu (Cheops), who preceded Khafre. In fact, this stela credits Khufu with repairing the Sphinx after it was struck by lightning—which is consistent with the obvious repair work done on the monument and the Sphinx and Valley temples. Modern Egyptologists generally treat the Inventory Stela as a late-period fabrication of an Old Kingdom text, and refuse to accept it as authentic.

Selim Hassan examined the literary references of the ancients to the Great Sphinx in the period stretching from the New Kingdom to Roman times, and found that all of them considered the Sphinx older than the pyramids. Interestingly, the oral traditions of some of the villagers who live around Giza agree that the Sphinx is at least five thousand years older than Khafre.

In short, there is circumstantial evidence on both sides, some indicating that the Great Sphinx may date to the time of Khafre, the rest indicating that it is older. The physical evidence is the deciding factor: The Sphinx had already been in place for at least 2,500 years when Khafre laid the first block of his pyramid.

Setting the Firestorm

By October 1991, I felt confident enough about my research to present it to the annual meeting of the Geological Society of America. Geologists, I find, are a forthright and honest bunch of scientists. I knew that if I had misinterpreted any of the data, somebody would point out the error of my ways to me. It greatly relieved me that no one found any flaws in the work. Actually, a remarkable number of my colleagues found the project so interesting that they asked to be included in the research if I needed help.

Then the trouble started. The popular press, which attends scientific meetings on the prowl for interesting news, picked up on the story and put it out. To my great surprise, articles on the Sphinx research soon appeared in *The New York Times, The Washington Post, The Independent* of London, even *USA Today.* A number of Egyptologists, archaeologists, and geo-archaeologists took immediate, sometimes angry exception to my hypothesis that the Sphinx was older than they thought.

The hoopla led to an invitation to debate the age of the Sphinx under the aegis of the American Association for the Advancement of Science, in February 1992. It wasn't much of a debate, actually, but a series of short talks with no time for discussion. Thomas Dobecki and I were on one side, and Mark Lehner and the geologist K. Lal Gauri were on the other. The press continued the story, reporting on the high drama of an academic controversy fanned by Lehner's tendency to indulge in name-calling and insult.

As a scientist, though, I needed to pay close attention to what the other side was saying. After all, there is always that small worry lurking in the back of a scientist's mind that something critical may have been overlooked. I have gone through all the countervailing arguments raised in the ongoing controversy, to see whether any of them contains that critical missing detail. So far, I haven't found it.

Farouk El-Baz, who is director of Boston University's Center for Remote Sensing, maintains that the Great Sphinx is in fact a yardang—that is, a natural, wind-shaped hill—which Old Kingdom artists simply retouched to make it look the way it does. Thus, he argues, the rain erosion I documented happened to an exposed rock outcropping long before any ancient Egyptian sculptor ever picked up a chisel to transform a piece of nature into a work of art.

But El-Baz's idea just doesn't fly. The Sphinx lies below the surface of the Giza Plateau, and the ancient Egyptians had to dig a trench around it to expose the stone. Before that excavation, the limestone was covered over by surface layers of sand or soil. The weathering could have begun only after the ancient Egyptians excavated the enclosure to carve the statue. The head, it is true, may have been a yardang, but it has been so heavily reworked that it's now impossible to tell for sure.

Frank J. Yurco, an Egyptologist at Chicago's Field Museum of Natural History, takes a different tack. He argues that the body of the Sphinx contains much poorer limestone than the other monuments at Giza. As a result, the Sphinx cannot be compared with the other structures because, under the same climatic conditions, the Sphinx would have weathered much more severely. He says, too, that since the Sphinx is low-lying, it was flooded repeatedly by the rising Nile. In other words, water did damage the Sphinx, but the water came from the river, not the rain-filled skies of the Nabtian Pluvial period.

Yurco's ideas sound superficially plausible, but they don't stand up to scrutiny. He's right about the low quality of the limestone. Still, as far as we can tell, this same limestone was used throughout much of the Giza complex, and can be expected to weather at the same rate and in the same manner over the same period. Because the limestone is relatively homogeneous, more weathering indicates the passage of more time, not nonexistent differences in the rock. As for flooding, the Nile did indeed reach the Sphinx, at least on occasion. Yet if inundation was the cause of the water-induced erosion, the wear should be greater on the paws and lower body, when in fact it is most pronounced on the much higher back. Damage of that sort could have come from flooding only if the statue was awash to its neck again and again in floods that would have given Noah pause. How could that have happened when the historical record says clearly that the Sphinx was buried in sand for much of the period between 2500 B.C. and 1400 B.C.? Additionally, Yurco's argument doesn't account for the differences in weathering in the rocks of the Sphinx enclosure. The entire floor of the enclosure lies at the same elevation, and any flood large enough to wet part of it would wet all of it. How, then, can it be that the east, north, and south sides are so much more deeply weathered than the west?

Kathryn Bard and George Rapp Jr., both of Boston University's Archaeology Department, take a similar bad-rock approach, arguing that the Sphinx isn't a normal massive limestone formation. They also make much of the fact that it still rains in Egypt, which complicates the weathering picture.

Actually, there is no such thing as a "normal" limestone, and the properties of Giza's limestone are well studied. I took that work into account in developing my hypothesis. And, of course, as Bard and Rapp state, it still rains in Egypt. The issue, however, is that these days it rains much less than it did during the Nabtian Pluvial of predynastic times. Only that kind of long-gone, consistent, heavy rainfall accounts for the Sphinx's obvious precipitation-induced weathering and the deep erosion in the surrounding enclosure.

One geologist suggested to me that the apparent deeper

weathering on the east, north, and south sides of the Sphinx enclosure resulted not from earlier exposure to the elements, but from the southeasterly dip in the rock strata of the underlying formations. The differential weathering pattern, however, doesn't follow the dip of the strata. Instead, it parallels the floor of the enclosure and actually cuts across the dipping strata. Additionally, the weathering of the east, north, and south sides is remarkably uniform despite the dipping strata. It appears nearly certain that the wear resulted from exposure of the rock by human activity, not from differences in the subsurface strata.

K. Lal Gauri, a geologist at the University of Louisville, has made much of the "rapid" weathering of exposed limestone. He argues that stone of this sort weathers away much faster than I have allowed for, and that my analysis mixes and matches different strata of limestone. Gauri's second argument is simply wrong. My research was careful in analyzing different weathering rates only in limestones of identical strata. Geologically, I made sure to compare apples with apples.

As for Gauri's argument regarding rapid weathering, it contains a peculiar irony: The conventional position holding for a 2500 B.C. origin of the Sphinx requires a much faster rate of weathering, under conditions that normally do not cause limestone to weather very quickly, than my hypothesis does. As I pointed out earlier, the Sphinx had worn so severely that it needed repair even in pharaonic times. It is plausible that the Sphinx could have weathered from a freshly carved state to its current condition in 4,500 years if it had never been repaired, restored, or buried under a protective layer of sand. The question is whether it could have become so severely weathered in ancient times alone, before the first repairs were made. Most Egyptologists are in agreement that this repair effort took place no later than 1400 B.C., with a few arguing for an earlier date of around 2200 or 2300 B.C. If they are right, and if the Sphinx was first constructed in 2500 B.C., then it had to weather severely in only 200 to 1,100 years, during a period when it is known to have been buried in sand for much of the time and when the climate was every bit as bone-dry and arid as it is today—conditions that tend to preserve limestone, not wash it away. This is indeed a much less likely model than a Sphinx that was built between 7,000 and 9,000 years ago and eroded during the heavier rains of that earlier era, long before repair was required at least thousands of years later.

In the end, the physical evidence stands. Based on what we now know, the Great Sphinx of Giza was constructed somewhere between 7000 and 5000 B.C.

Upsetting the Apple Cart

The arguments advanced to support the 2500 B.C. dating of the Sphinx and attribute it to Khafre are ad hoc ideas invented to protect an established chronology. It is as if the Egyptologists scratch their heads and say, "But we *know* the date's got to be 2500 B.C. Surely Schoch is wrong. He simply has to be." Unwittingly, they are caught in the trap of defending their own dogma rather than examining the evidence with the dispassion it deserves.

Of course, as a geologist, I come to Egyptology as an outsider. By and large, Egyptologists don't use this kind of scientific evidence; they rely instead on a mix of methods that includes historiography, archaeology, anthropology, philology, and literary analysis. Geological analysis is an alien form of thinking, one that Egyptologists are likely to reject because of their own lack of familiarity with it.

Yet lying even deeper under the terms of this debate is an overriding assumption about the capacities and capabilities of ancient peoples. Mark Lehner told the *New York Times,* "If the Sphinx was built by an earlier culture, where is the evidence of that civilization? Where are the pottery shards? People during that time were hunters and gatherers. They didn't build cities." Carol Redmont of the University of California, Berkeley, made a similar statement to the *Los Angeles Times:* "There's just no way that could be true [that the oldest portion of the Sphinx dates back to 5000 B.C. or earlier]....The people of that region would not have had the technology, the governing institutions, or even the will to build such a structure thousands of years before Khafre's reign.... [It] flies in the face of everything we know about ancient Egypt."

Does it really? Or does it merely undermine our assumptions? Were the people of long ago unlettered knuckle-draggers who didn't know what a sphinx was, much less how to build such a massive example of one? Or have we, in the assumed superiority of our own culture, arrogantly written off an ancient, vanished people different from ourselves yet sophisticated in ways we do not understand?

Ancient Origin: Civilization's Rescheduled Beginning

RESPONSIBILITY FOR THE IDEA OF EVOLUTIONARY PROGRESS IS often laid unfairly at the feet of Charles Darwin. In fact, the notion reaches much farther back than the father of the theory of evolution by natural selection. Admittedly, Darwin's theory did lend itself to the idea that as species change over time, they inevitably become better and better. The concept of natural progress over the years, centuries, and millennia was very much a feature of the European Enlightenment in the eighteenth century, a hundred years before Darwin. Indeed, its roots extend to an even earlier time, at least to the early centuries of the Christian era. The old religions of the Greek and Roman classical world saw history as recurring cycles of birth and decay. They portrayed the passage of time as a circle: what goes around comes around, again and again. Christianity straightened history into a line, reaching from Moses to Jesus to Judgment Day in an inevitable and divine progression.

Given that the scholarly and scientific cultures of the United States and Europe are very much the products of both the Enlightenment and the Christian worldview, the notion of necessary improvement over time has subtly but indelibly influenced our paradigm of civilization's rise. The existing model of prehistory posits steady progress step by step from the first group of Stone Age hunters trying their hands at farming to the bustling urbanites of today's Manhattan.

According to this model, during the so-called Paleolithic period from modern humankind's earliest origins until about 8000 B.C., we lived by hunting wild game and foraging for roots, fruits, and other plant foods, moving from place to place in the endless nomadic pursuit of a full larder. Around 8000 B.C., particularly in the Fertile Crescent of the Near East, some of these nomads settled into permanent villages, sparking what has been called the Neolithic revolution. Because they had learned how to sow and reap grain, and had mastered the art of domesticating and raising sheep, goats, and cattle for milk, meat, wool, and hides, these people no longer needed to move with the seasons. As humans became increasingly adept at farming and herding and fully exploited the opportunities offered by this new settled economy of agriculture and herding, villages grew into towns, towns into small cities. By circa 3500 B.C., the world's first true civilizations had arisen more or less simultaneously in Mesopotamia and Egypt, ushering in the Bronze Age with its metal weaponry, full-blown writing systems, and complex social, political, and religious organizations.

My research on redating the Sphinx caused a stir not only because it challenged the traditional chronology of orthodox Egyptology, but also because it undercut this model of steady progress from Paleolithic to Neolithic to Bronze Age. If the Great Sphinx of Giza dates to the 7000–5000 B.C. period, it should be Neolithic. But what uncivilized Neolithic folk—people who, according to the conventional model, tended flocks and herds, defended themselves with stone-tipped arrows and spears, farmed crudely on small plots, and clothed their bodies in the uncured hides of animals—could have carved so magnificent a sculpture in such colossal proportions? The Sphinx is a triumph of artistic genius and of both organizational and technical know-how, qualities long assumed to be light-years ahead of the capacity of the Neolithic Age and to have been unattainable before the advent of the Bronze Age.

Yet the anomaly remains. And, as we shall soon see, it is only one of the many pieces of a complex puzzle whose assembled picture shows that a linear, progressive vision of the rise of civilization is an illusion, and that our cultural roots reach back to a far earlier time than we have previously thought possible.

An Egypt Swept Clean

When most Egyptologists are presented with a hypothetical dating of the Great Sphinx to the 7000–5000 B.C. period, they point immediately to the lack of physical evidence indicating the presence of a civilization capable of supporting such an effort. That statement is true, as far as it goes. Very little archaeological evidence from the end of the Paleolithic period to the beginning of the Bronze Age has been unearthed in the Nile Valley and Delta. On the face of it, this limited evidence could indicate that few humans inhabited these regions during that period. Yet it must always be remembered that an absence of evidence is not evidence of absence.

Most of the archaeological material from the predynastic period of Egypt comes not from the river's cultivated floodplain but from habitation sites and cemeteries, like the Giza Plateau, that lie at the edge of the desert or along the southern edge of the Nile Delta and sit at some elevation above the lower-lying flats. Very little archaeological excavation work has been carried out in the delta itself or in the valley, and for good reason: extreme difficulty. In the days before the Nile was dammed, the river's regular flooding deposited an average of one millimeter of alluvial soil each year across the delta and the rest of the river's floodplain. Those deposits added up. In the past 10,000 years, the span between the traditional beginning of the Neolithic period and today, the Nile Delta and Valley have been progressively buried under eight meters, or a little more than twenty-six feet, of deposited soil. Removing an overburden of that depth poses great technical difficulty. As a rule, archaeologists like their ruins better exposed, closer to the surface; it helps to know where to begin digging.

To complicate matters, the bed of the Nile in its lower reaches has shifted over the millennia, so that what is riverbank now may have been underwater at various times thousands of years in the past. Additionally, much of what was inhabitable coastline in Egypt several millennia ago now lies underwater. The retreat of the last ice age launched not only the heavy rains of the Nabtian Pluvial, but also raised sea levels profoundly and quickly. Beginning around 8000 B.C., the Mediterranean rose an estimated 200 or more feet, burying beneath its waters any villages, cities, or religious sites used by the dwellers on what used to be the coastline.

The two Neolithic settlements known from Lower Egypt in the fifth millennium B.C. are less sophisticated than one would expect for the period. These were the ragged camps of rough, tough people eking out a hardscrabble existence on the edge of starvation, the kind of place where physical survival rather than high culture was the first order of business. How could such a backwater have been the source of the civilization that built the pyramids?

Prehistorian Mary Settegast doubts this is the case. "It has never seemed logical that the Nile Valley would be almost uninhabited during a period when lands to the east and west of Egypt were experiencing great advances in population and cultural development," she writes. "Moreover, the relative backwardness of the two Neolithic settlements which then do appear in fifth-millennium Northern Egypt (Faiyum and Merimde) is not what one expects of the times, and several archaeologists now suspect that both may actually have been marginal settlements, perhaps of Libyan bedouin..., rather than true representatives of the cultural level of fifth-millennium Lower Egypt."

In fact, other evidence from ancient Egypt, though scant, adds to the notion that the remnants of predynastic culture lies undiscovered beneath the waters of the Mediterranean or the accumulated soil of thousands of Nile floods. An intriguing example is the so-called Libyan palette, a predynastic (circa 3100–3000 B.C.) artwork on display in the Egyptian Museum in Cairo. The palette shows seven structures that look like fortified cities located along the western edge of the Nile Delta. If we take this palette as credible evidence—and there is no reason not to—it clearly indicates the existence of cities in predynastic times. However, no such cities have been uncovered. Very possibly they lie buried under the shifting desert sands, the accumulated Nile silt, or the invading waters of the Mediterranean.

On a cultural and technological level, a great deal has been going on in Egypt for a very long while. As long ago as 33,000 years, the inhabitants of Upper Egypt displayed the significant technical and organizational skills needed to mine stone. P. M. Vermeersch and his associates excavated a shafted, chambered flint mine that is radiocarbon-dated to 31,000 B.C., making it the oldest such mine yet discovered. Early Neolithic sites dating to circa 8100 B.C. in Egypt's Western Desert feature well-planned, sophisticated villages complete with wells, habitations much more sophisticated than the primitive fifth-millennium sites.

The most exciting pertinent discovery about predynastic Egypt is very recent, announced only in March 1998. Several years ago, a team led by Southern Methodist University anthropologist Fred Wendorf located a complicated Neolithic ruin in the Nubian Desert of the southern Sahara. The site lies on the edge of what used to be an ancient lake, which began filling with water in about 9000 B.C. as a result of the increased rainfall of the Nabtian Pluvial, and remained full until rainfall fell off after circa 3000 B.C. Nomadic cattle-herding peoples used the area to graze their animals on a seasonal basis from circa 8000 B.C. until the lake dried up in approximately 2800 B.C., making the area no longer habitable. Using huge stones, or megaliths, the nomads populating Nabta during the wet times built a stone circle, a number of flat, tomblike structures, five lines of standing stones, and villages with deeply dug, walk-in wells. A number of the tombs were apparently shrines to cattle; one of them contained a fully articulated bovine skeleton, and another held a rock sculpted to resemble a cow. Apparently these unknown people used cattle in their religious rites much as the nomadic, cattle-herding Masai of East Africa do today.

The most fascinating aspect of Nabta is less the cattle worship than the circle and alignments of standing slabs. Follow-up work at the site by Wendorf, University of Colorado astronomer J. McKim Malville, Ali A. Mazar of the Egyptian Geological Survey, and Romauld Schild of the Polish Academy of Sciences made use of satellite surveys and showed that one of the lines runs exactly east to west, altogether too perfectly to be a coincidence. The stone circle contains four sets of stone slabs; two sets are aligned north to south, and the second pair provides a line of sight to the horizon where the summer solstice sun rose about 6,000 years ago.

Nabta could also have been used to mark the zenith sun. Since the site lies just south of the Tropic of Cancer, the noon sun reaches its highest point in the sky on two specific days, one three weeks before the summer solstice and the other three weeks after. On those days, upright objects like the standing stones of the circle and alignments cast no shadows. The zenith sun has been a major event for many tropical cultures for millennia, and the stone alignments allowed Nabta's inhabitants to time the event precisely.

These long-ago people went to a great deal of trouble to achieve this precision. Some of the slabs reach nine feet high, and they were dragged to the site from a sandstone outcrop a mile or more away. That effort took quarrying skill and enough organization to assemble a team to cut and move the heavy, dangerous slabs and put them in place.

Of course, astronomical alignments of this sort are known from various other places in the world, the most famous being England's Stonehenge and similar sites in Brittany, Ireland, and other parts of Europe. Nabta, however, is a full millennium older, dating to between 4500 and 4000 B.C., and it qualifies as the most ancient astronomical alignment yet discovered. The people who built it possessed the organizational and technical skill to erect such monuments, and they made use of a highly detailed astronomical knowledge of the sun's movements.

The people of Nabta were not unique in their skill and knowledge. When we move the circle of inquiry outside Egypt to the neighboring lands of the Near East, pronounced sophistication announces itself earlier and earlier. Consider the ancient city of Jericho, in what is now the Israeli West Bank, lying about 200 miles east of the Nile Delta. Dating back to circa 8300 B.C., Jericho boasted a massive stone wall that was 6.5 feet thick and at least twenty feet tall. Outside this wall, a ditch twenty-seven feet wide and nine feet deep was cut into the solid bedrock, probably to serve as a moat. Within the protection of the wall, Jericho's ancient inhabitants built a stone tower thirty feet in diameter and at least thirty feet tall—that's the height of the ruins today; the original may have reached even higher—with a flight of steps fashioned from huge stone slabs running up its center. Apparently perceiving a threat from the outside and needing to protect themselves, the people of Jericho had the organizational and technical skill to create a defensive structure that compares favorably to the feudal siege castles of medieval Europe built over 9,000 years later.

Farther north, in the Anatolian region of what is now Turkey, lie the ruins of the ancient city of Çatal Hüyük, which are dated to the seventh millennium B.C. Compared to Jericho, which is estimated to have contained only several hundred residents,

Çatal Hüyük was a huge, bustling place, with 7,000 inhabitants living in an orderly complex of mudbrick-and-timber houses and temples. Religious and symbolic life was developed and rich, expressed in ornate wall paintings and sculptures depicting bulls' heads, female breasts, frogs, and vultures, images thought to represent the forces of birth, death, and regeneration.

Nabta, Jericho, and Çatal Hüyük all display evidence of people able to organize themselves to accomplish complex tasks and possessed of impressive artistic, technical, and engineering knowledge. And none of these sites is a prototype, the sort of first-time experiment, half success and half failure, that betrays the hand of a talented beginner with both great vision and rank inexperience. These sites were the work of people who knew what they were doing because they had been doing it for a long while, far longer than the old model of the rise of civilization allows.

Pushing Time Back in the New World

The sophistication of the ancient Near Eastern sites, including the Great Sphinx of Giza, shows that many of civilization's elements were in place for millennia before the accepted date of civilization's arrival, in 3500 B.C. The horizon of history needs to be rolled back to a time thousands of years earlier than we have thought possible. And this fact holds true not only for the Old World cradle of civilization, but also for the New World of the Americas.

Until well into the mid-1990s, the prevailing model held that humans first came to the New World from Asia by crossing a land bridge over what is now the Bering Sea between Siberia and Alaska. But for millennia, anyone who made it across the land bridge had nowhere to go. About 20,000 years ago, at the outset of the most recent ice age, two massive, thick ice sheets joined over what is now Canada and the northern United States, foreclosing any possibility of passage south. Then, some 7,000 years later, with the ice age beginning to come to an end in circa

11,000 B.C., an ice-free corridor opened between the glaciers, providing a southern route. Anthropologists and archaeologists theorized that the first Americans making their way across the land bridge at this time migrated south as the ice retreated, giving a clear route into a land that to them was a new opportunity. This hypothesis fit with the evidence of what was thought to be the oldest discovered site of human occupation outside Alaska, at Clovis, New Mexico. Clovis showed that humans in the area were hunting and butchering mammoths during the centuries between 9500 and 9000 B.C. That date worked well. It was more recent than the oldest known sites in Alaska—after all, if the migrants moved north to south, then northern sites had to be older than Clovis, as indeed they proved to be—yet it allowed enough time for the new arrivals to cover the considerable distance in between.

Various scholars made claims of discovering human habitations older than Clovis, but again and again the claims were disproved. From its discovery in the 1930s through the 1990s, Clovis remained the benchmark for the presence of humankind in the New World.

Until, that is, the discovery of an ancient campsite at Monte Verde, in southern Chile. Twenty years of research by a team of American and Chilean scientists, led by the University of Kentucky's Tom D. Dillehay, have swept away the last skepticism among scholars and shown that Monte Verde dates to 10,500 B.C., at least 1,000 years before Clovis. Such a small change—after all, in prehistorical terms, what's a millennium more or less?—might seem to be no big deal, yet it utterly upsets the former model. Monte Verde lies 10,000 miles south of the Bering land bridge, separated from it by vast tracts of mountain, plain, and jungle and climate zones ranging from arctic to tropical. There is simply no possibility that Paleolithic hunters crossing over the Bering Sea land bridge when the ice-free corridor opened in circa 11,000 B.C. could have traveled such an immense distance, either on foot or using small boats to come down the coast, in a mere five hundred years. To have covered an expanse of this order, the ancestors of Monte Verde's inhabitants had to have entered the New World before the ice sheets converged in circa 18,000 B.C.

Other research indicates that even this date is perhaps far too recent. At the Monte Verde site, Dillehay's group found charcoal in a probable fire pit that was radiocarbon-dated to 31,000 B.C. Recent discovery of a Paleoindian site in Brazil near the Amazon River that is contemporary with Clovis but culturally distinct suggests a far longer background of cultural change than the old model allows. And Johanna Nichols, a linguist at the University of California at Berkeley, has calculated that development of the more than 140 languages spoken by the native peoples of the Americas would have required at least 30,000 years—a date interestingly consistent with the Monte Verde charcoal.

Whatever the exact date of humankind's first entry into the Americas, the New World isn't as new as we thought. Humans have been here much longer than we realized, and our history on these two continents stretches back to an earlier time than we had considered possible.

The Star Bulls

Of course, there is more to the redating of the Sphinx than a matter of moving the horizon of history, in both Old and New worlds, farther back into time. We also need to look for evidence that these increasingly ancient peoples possessed the kind of sophisticated knowledge we now assign to the realm of science. As we learn more about the ancient peoples who lived before what we think of as the dawn of civilization, we find again and again, as in Çatal Hüyük and Jericho, that they were more skillful than we thought possible. Another example, one that demonstrates a striking knowledge of astronomy among Paleolithic humans, can be found in the south of France.

The discovery of the cave of Lascaux in the Dordogne Valley in

1940 shattered any smugly modern feeling that these people of 17,000 years ago lacked esthetic sensibility or artistic skill. The vast, multichambered cave was decorated with exquisite friezes of animals and strange, shaman-like creatures, part human and part beast. Modern painters studied the pictures to learn the insights of this long-lost art, while archaeologists puzzled over the paintings' meaning. Obviously, Lascaux qualified as a religious site, but of what sort? The scholarly consensus was that Lascaux served as a temple whose paintings of animals ensured success in the hunt. This Sistine Chapel of the Paleolithic depicted an ancient people's happy hunting ground.

The most magnificent of the Lascaux friezes is the Hall of Bulls. On the cave's domed ceiling, huge bulls leap and prance, a herd of stags swims a stream, bison shake their horned heads, and woolly ponies, many of the mares heavy-bellied with unborn colts, trot on delicate hooves. One of the creatures is particularly magical, a beast that never existed. Sometimes it is called the Unicorn, a name that doesn't really fit. Two long, straight horns, not one, point from its horselike head, while its bulging belly, which nearly touches the ground, and humped upper back lack the grace usually attributed to unicorns.

In circa 15,000 B.C., when the Hall of Bulls was painted, Lascaux was in the grip of the last ice age and much colder than it is today. The principal game animal hunted by the people living in Dordogne Valley at that time was reindeer. This fact makes it difficult to understand the Hall of Bulls as a magical happy hunting ground. These people hunted reindeer. Why, then, did they put such effort into painting bulls they did not kill, and creating some strangely horned or antlered creature that no hunter had ever encountered or sought?

A fascinating and cogent explanation has been offered by Frank Edge, a community college and high school teacher of mathematics and cosmology. The bulls provide not hunting magic, according to Edge, but a map to the summer sky of 17,000 years ago.

The key to Edge's understanding arises in a striking resemblance between seven dots painted over the shoulder of one of the Lascaux bulls and the star cluster known as the Pleiades. The face of this bull also displays a series of facial dots that closely resemble the star Aldebaran and the Hyades star field. The Pleiades, Aldebaran, and the Hyades all fall within the constellation we moderns call Taurus—that is, the Bull.

"So striking is the resemblance of this ice age bull to the traditional picture of Taurus," Edge writes, "that if the Lascaux bull had been discovered in a medieval manuscript rather than on a cave ceiling, the image would immediately have been recognized as Taurus."

The Lascaux frieze wasn't painted as a single original composition. Stylistic differences and the pigments used indicate that the bulls, one horse's head, and the Unicorn were all composed in black-manganese paints by a single hand or a small group of painters working in collaboration. The red cows had already been painted on the cave roof prior to that time, and the many-colored ponies were added afterwards, fitted into the spaces left around the large black-manganese animals.

Beginning with the bull of the Taurus constellation, Edge turned his attention to the other black-manganese paintings, looking for celestial clues in these depictions. He found them. Bull after bull provided keys to the prominent stars of the summer sky. Edge determined, for example, that the Unicorn's strange shape came from combining the stars in the constellations we know as Scorpio, Libra, and Sagittarius into one super-constellation. For example, the Unicorn's two horns match the two bright stars of Libra, the humped back replicates the Antares arch in Scorpio, and the oddly sagging belly corresponds to the curve in Scorpio's tail.

When all the black-manganese figures in the hall are assembled into a single composition, they provide a remarkably accurate picture of the ring of the summer night sky of 15,000 B.C. Someone who stood outside and marked the position of the prominent stars and then stepped inside the cave would see the same image re-created in the bulls and the horse's head.

But why? For what reason would these ancient artists have gone to such trouble to depict a particular night sky?

The answer, Edge suggests, is that it allowed them to fix precisely the date of the summer solstice by observing the path of the moon. During the summer—the season when the cave was in use, as revealed by pollen samples found inside—the moon passed through the star pictures drawn on the ceiling. Indeed, only in the very months near the summer solstice did the moon travel through all of the star clusters depicted by the bulls, Unicorn, and horse's head.

"The full moon prior and closest to the summer solstice would stand, each year, shining from the region of sky between the faces of the two opposing bulls...," Edge writes. "The spring full moons, prior to the solstice, would stand among the left-facing figures, while the full moons following the solstice would stand among the right-facing figures....Thus we see that with a knowledge of the moon phases related to the star figures portrayed in the Hall of Bulls, a Lascaux observer could predict the coming of the summer solstice and roughly calculate the time before or after that event."

In the pre-Christian religions of Europe, the summer solstice, or Midsummer's Night, was a key religious festival. Clearly the roots of this observance reach far back, to the people of Lascaux and beyond. No doubt, too, this knowledge had been in existence for some time, for Lascaux, like Jericho or Çatal Hüyük, is no first-time effort. Working from long experience, these ancient artist-astronomers knew exactly what they were up to.

Part of the beauty of the Lascaux system is its striking simplicity, as Edge notes. Using the cave picture as a star map to fix the summer solstice "required only the careful observation of the moon phases with respect to the celestial animals of spring and summer. The system could be taught to anyone and carried in memory without writing."

As intellectually powerful as this system is, it was already ancient when the painters of Lascaux made use of it. According to the work of Alexander Marshack, people in the Dordogne were observing the phases of the moon at least 15,000 years before the paintings were made—that's 30,000 B.C. Marshack has also found evidence that within five millennia after the Lascaux paintings, inhabitants of the same area had developed a non-arithmetic calendar of the sun and moon that allowed them to determine the time of the solstice independent of the movement of the moon. Like Edge, Marshack is convinced that we moderns underestimate the sophistication of ancient people, who were in fact much more advanced culturally and intellectually than we give them credit for. Lascaux provides persuasive evidence of an ancient people who were artistically adept and intellectually capable of ascertaining the recondite movements of the celestial bodies.

Another fascinating aspect of Edge's work is the persistence of celestial images over unexpectedly long reaches of time. Constellations exist, of course, in the eye of the beholder, not in the sky itself; it takes one trained to look for particular images in the starry sky to find those very images. Evidence suggests that the three prominent star groups we know as Leo, Taurus, and Scorpio received their names by 4000 B.C. and were used as markers for the beginnings and ends of seasons. Edge's research shows that the image of the bull, the constellation Taurus, is even older.

Edge concludes, "[T]he suggestion that any one celestial/mythological image has been with us unchanged for 17,000 years (nearly 1,000 generations) emphasizes the great importance of the heavens in framing the mythology of the Western world, while also offering an outstanding example of the power and endurance of at least one oral mythological tradition."

Alexander Marshack agrees. In his view, the people of the Upper Paleolithic "had a level of observation and story use, which could, when faced with the climate and ecological changes of the postglacial period, help them in developing specialized economies and mythologies in various areas and under specific local conditions."

By turning now to the mythological stories told about the starry heavens since deepest antiquity, we can come to understand even more about the intellectual sophistication of these long-ago people and their striking powers of observation.

Precession's Slow Turn

Ancient peoples not only knew the accurate positions of the stars in the heavens on particular dates, but also understood that the pattern of stars and constellations moved in a most subtle fashion because of the phenomenon known as precession, which is caused by a slow wobbling of Earth on its axis.

Our planet isn't really round; it flattens at the poles and bulges at the equator, so that a radius drawn from Earth's center to the equator is 13.5 miles longer than one drawn to either pole. This extra mass in the middle makes Earth less a sphere than what is known technically as an oblate spheroid. Additionally, Earth's axis of rotation tips in relation to the plane (also known as the ecliptic) of its orbit around the sun. The sun and the moon and, to a much lesser extent, the other planets, tug gravitationally on the greater mass of the tipped Earth's equatorial bulge and, because of this slightly imbalanced pull on the planet, slowly move the axis of rotation. As a result, Earth spins not like a wheel on an axle, round and round in the same plane, but with the wobble of a top moving across a table or a floor.

This slow wobbling movement, known as precession, affects astronomical events viewed from Earth. For example, the celestial north and south poles are not fixed and permanent; instead, they shift slowly in relation to the positions of the stars. Currently, as every Boy Scout studying wilderness navigation learns, the celestial north pole is marked by the star called Polaris (also known as Alpha Ursae Minoris), at the tip of the constellation Ursa Minor. Polaris won't always be the North Star, however, nor has it always been. Because of precession, the celestial pole traces a circular path through the heavens, completing a full cycle approximately every 26,000 years. In 12,000 B.C. the North Star was Vega. In 3000 B.C., during the beginning centuries of dynastic Egypt, the North Star was Alpha Draconis. At the height of ancient Greek civilization in the fifth and sixth centuries B.C., it was Beta Ursae Minoris. Already the celestial north pole is moving away from Polaris, and by about 14,000 A.D., Vega will again become the new North Star guiding night-hiking Boy Scouts and completing this most recent round of the cycle of precession.

Precession affects the celestial equator as well as the poles, very gradually moving the whole vault of the heavens in relation to the eyes of an Earth-based observer. Over time, constellations rise at new points on the eastern horizon and follow different paths across the sky to set in the west. And it affects, too, the position of the sun relative to the stars, a phenomenon that was particularly noted on the important ritual days of spring and fall equinox, around March 20–21 and September 22–23, when the sun is positioned directly over the equator. On the spring equinox in our era, the sun rises against the constellation of Pisces. Three thousand years ago, it was rising in the constellation Aries.

With its cycle of approximately 26,000 years, precession can hardly be observed in the span of a single lifetime. Detecting this long, subtle motion required long, careful observation of the stars and recording of their positions over centuries. The honor for the discovery of precession is conventionally credited to Hipparchus, a brilliant Hellenistic mathematician and astronomer of the second century B.C. Watching the sky one night, Hipparchus was surprised to see a star where he was certain no star had appeared before. He then carefully cataloged each of the 1,080 fixed stars he could see, positioning them by celestial latitude and longitude, then compared his sky chart with one made by an earlier Greek astronomer a little over a century and a half before. Calculating that all the stars had shifted position by approximately two degrees in the intervening years, Hipparchus named the phenomenon precession.

But did Hipparchus actually discover precession? He may have drawn from Babylonian astronomical records, particularly those compiled by Kidenas (in cuneiform, Kidinnu) in the fourth century B.C., which had come to the attention of Greek scholars after the conquest of the East by Alexander the Great. But there is tantalizing and persuasive evidence that precession was known long before the Babylonians, long before history.

Time's Terrible Mill

Mythology shows that humankind in ancient times understood much more about the celestial realm, including precession, than we realize. This statement may seem surprising, for we moderns fail to take myth seriously and thus miss the information it contains. Indeed, we use the word *myth* as a synonym for "lie" or "falsehood," revealing an unfortunate ignorance of this marvelous store of beauty and wisdom.

Myth creates a frame of understanding largely alien to the modern world's mind-set. Let's suppose that you or I want to describe the blinding speed with which a particular athlete runs. In a technological or scientific state of mind like that of our contemporary civilization, we state this fact by providing a quantitative measurement for his speed: "Athlete X runs the hundred-yard dash in 9.9 seconds." A poet invents a simile: "He runs like a deer." The mythmaker goes further, equating athlete with deer: "He is a deer." Strictly speaking, of course, the mythic metaphor is false. The athlete's not a deer, but a man, a very fast man, a man so blazing that his speed gives him something of the essential qualities of a deer, making the two beings both impossibly and wonderfully the same. This is myth, a way of making connections in our universe at a level deeper than the superficially obvious.

In the ancient world, myth was highly valued, both for its efficiency and its complexity. Mythic stories are vivid, colorful, and laced with intrigue, strange beasts, sexual variety, blood, guts, hero-feats, and more than enough soap-opera treason, infidelity, dalliance, and conflict to make them easy to remember and exciting to tell. This mnemonic aspect of mythology was critical in an ancient world that lacked writing, where memory carried the burden of preserving all knowledge. Scholarship has shown, for example, that both Homeric poems, the *Iliad* and the *Odyssey,* existed for centuries in oral form before they were written down. Poets learned the tens of thousands of lines in each epic by heart, then recited them from memory. Poetic devices like the dactylic hexameter and repeated phrases such as "the wine-dark sea" and "gray-eyed Athena" served not only esthetic ends—they are part of the reason why these poems sound beautiful and pleasing to the ear even today, 3,000 years after they were composed—but they also eased the task of memory, making it simpler for the poet to keep this huge body of mythological knowledge straight. The *Iliad* lends itself much better to memorization than does, say, *On the Origin of Species,* in large measure because of its epic-poetic structure and diction. But, like the *Origin of Species,* myths such as the *Iliad* were often freighted with knowledge. They conveyed science, philosophy, or history in the vivid tales used by the ancients rather than in the dispassionate discourse preferred by a scientific civilization such as ours.

Just in the past two decades, mythology has reemerged as an important area of study, inquiry, and even popular interest, largely as a result of the writings of Joseph Campbell, who was strongly influenced by the comparative-religion scholar Mircea Eliade and the psychotherapist Carl Gustav Jung. Campbell, Eliade, and Jung all viewed myth as a detailed road map to the struggle of the individual soul toward independence and transcendence. Seeing myth as spiritual in the broadest sense, they largely ignored its role as a record of human history or as a repository of knowledge about matters other than those pertaining to the soul's journey.

There have been dissenters to this point of view. A significant example is Robert Graves, the English poet and writer. In several brilliant books on Greek and Hebrew mythology, Graves highlighted the historical themes behind many of the ancient stories, and in *The White Goddess,* he depicted the Celtic mythology of Ireland and Wales as a mode of transmitting secret Druidic information from adept to initiate, generation to generation.

This second approach, one that looks at myth as a body of complex knowledge designed to be passed on in a preliterate culture, was used by Giorgio de Santillana, a philosopher and historian of science at the Massachusetts Institute of Technology, and Hertha von Dechend, a historian of science at the J. W. Goethe–Universität Frankfurt, to examine mythology's astronomical aspects. Their work resulted in a book titled *Hamlet's Mill,* which, though occasionally pedantic and often inaccessible to nonspecialists, is an eye-opener.

Joseph Campbell argued that a single, original mythic story, which he called the monomyth, underlies all the various ethnic and national versions of mythology and points to the beginnings of mythology at a single place and time in the prehistory of humankind. Santillana and Dechend agree. They argue that aspects of what we think of today as separate mythologies—Greek, Hebrew, Teutonic, Icelandic, Polynesian, Irish, and Native American, for example—are rooted in a single source of astronomical knowledge originating in the Middle East millennia ago, and that these stories show a clear understanding of the precessional cycle dating to thousands of years before Hipparchus.

Many mythologies open with the universe beginning in a perfect harmony that breaks down, often because of human wrongdoing. Dissension, disharmony, violence, and death enter the workings of the world, which slowly disintegrates and decays. This basic story appears in the one mythological system many Americans are still familiar with, the Bible. According to the Old Testament book of Genesis, the Garden of Eden was a place of abundance and harmony until Adam and Eve sinned by helping themselves to the fruit of a tree that God had forbidden them to sample. Everything came apart, the Garden was lost, and in no time at all the world witnessed its first murder, as the jealous Cain slew the dutiful Abel. This theme of an original perfection lost to a world that descends into increasing chaos can be found again and again in mythology, and points to the existence of the monomyth.

Another common theme concerns the transition from one world age to another, often amid cataclysm and catastrophe. Greek mythology is filled with world-ending images of titans and giants battling gods and goddesses, and the Norse stories are famed for their visions of *Götterdämmerung,* the twilight of the former, fading gods. Again, this kind of story is echoed in the Hebrew scriptures of the Old Testament, particularly in the flood that swept away all living things except Noah and his ark to cleanse the Earth for a new beginning. It is the core, too, of the Greek Christian scriptures of the New Testament, which proclaim a new age heralded by the birth of Jesus.

"What actually comes to an end is *a* world, in the sense of a world-age. The catastrophe cleans out the past, which is replaced by 'a new heaven and a new earth,' and ruled by a 'new' Pole star," Santillana and Dechend say. In other words, precession is the cosmological key to the historical shift from era to era.

Santillana and Dechend argue that the unknown philosophers of antiquity were obsessed with time. On Earth, time destroyed all things—people's lives, cities, kingdoms, dreams, and ambitions. Time and its endless mutability stood in contrast to eternity, a zone of perfection without time or change: "The true seat of immortality has always been denied to any aspect of 'time, the moving likeness of eternity,'" Santillana and Dechend write. "For eternity excludes motion."

The ancients could find the eternity they longed for just by looking up on a clear and cloudless night, where they beheld the stars fixed in their places turning in perfect circles across heaven's vault. Santillana and Dechend quote Aristotle: "'What is eternal...is circular, and what is circular is eternal.'" Myth concerned this eternal world. It created a new "Earth" in the sky, setting its tales within a geography of the heavens.

Yet, as the ancients discovered, the eternal only seemed eternal. Even the sky, they discovered, moved away from its ideal circularity and perfect lack of change. Precession, according to Santillana and Dechend, "was conceived as causing the rise and the cataclysmic fall of ages of the world." The ancients had noted that precession slowly moved the horizon position of sunrise on the spring equinox from one constellation to another. "The sun's position among the constellations at the vernal equinox was the counter that indicated the 'hours' of the precessional cycle—very long hours indeed, the equinoctial sun occupying each zodiacal constellation for about 2,200 years [actually the number is closer to 2,150 years]. The constellation that rose in the east just before the sun...marked the 'place' where the sun rested." This constellation in turn gave its name to the world age. In approximately 6500 B.C. the vernal equinoctial sun rose in the far eastern portion of Gemini, making the following two millennia the age of Gemini, as the equinoctial sun moved east to west against the stars of this constellation. Next came Taurus (circa 4300 B.C.), then Aries (circa 2150 B.C.), and finally Pisces (circa A.D. 1), where it will remain for a while longer before giving way to Aquarius in a little less than two hundred years.

Yet again, the Bible shows evidence of this system. Moses led the Hebrews out of Egypt after the vernal equinoctial sun had shifted from Taurus into Aries. Coming down from Mount Sinai, the Old Testament lawgiver is commonly depicted as two-horned, like Aries the ram, while the disobedient Hebrews who remained stuck in the old way worshiped the Golden Calf that symbolized the age of Taurus. And Jesus of Nazareth, born as the precessional hour shifted from Aries to Pisces (Latin for "fish"), is still commonly symbolized as a fish. Like Moses, Jesus is the herald of the new world-age symbolized by the movement of the precessional clock.

To the ancients, the truly important events in the universe occurred not on Earth, but in the sky. "Thus, the revolving heavens gave the key, the events of our globe receding into insignificance. Attention was focused on the supernal presences, away from the phenomenal chaos around us," Santillana and Dechend argue.

"The Precession," they continue later, "took on an overpowering significance. It became the vast impenetrable pattern of fate itself, with one world-age succeeding another, as the invisible pointer of the equinox slid along the signs, each age bringing with it the rise and downfall of astral configurations and rulerships, with their earthly consequences." In the ancient world, astronomy was a source of high anxiety.

The principal symbol for this notion of time grinding all before it is the millstone. Traced in Norse mythology to the hero Amlodhi—whom Shakespeare borrowed as the template for his tragic Hamlet, who does more than his fair share of fretting over the nature of life, death, and decay—the mill and the whirlpool it makes when buried in the sea appear in central roles in mythology after mythology. In the *Odyssey,* for example, Odysseus, who has returned to Ithaca and is ready to take back possession of his home, has a vision of a mill that will turn his enemies to dust. In the Old Testament book of Judges, the blinded, humiliated Samson is harnessed to a millstone to grind grain for his Philistine captors. And the central American god Quetzalcoatl creates a new race of humans by grinding up the bones of people killed in a great flood and adding divine blood to the grist.

As Santillana and Dechend show, mythology became a way of recording the events of the heavens, tracking the movement of precession and the grinding of time's mill, and providing a framework by which observers on Earth could determine what was truly happening over their heads. And it was done in a world without mathematics, without a writing system, without computers. Santillana and Dechend are right in calling this core of ancient mythology "an enormous intellectual achievement...in this organization of heaven, in making the constellations and in tracing the paths of the planets. Lofty and intricate theories grew to account for the motions of the cosmos. One would wonder about this obsessive concern with the stars and their motion, were it not the case that those early thinkers thought they had located the gods which rule the universe and with them also the destiny of the soul, down here and after death."

And how old is this system? In an aside, Santillana and Dechend suggest the monomyth's origin in about 4000 B.C., in the earliest days of the first known Sumerian astronomers. That is, I suggest, far too recent, by at least two millennia. The double ax, a symbol common in Knossan Crete and associated by Santillana and Dechend with the precession, is depicted on the walls of shrines in Çatal Hüyük that date to circa 6500 B.C. Another painting from the same period at Çatal Hüyük may be a mythological depiction of the movement of the seven planets among the twelve signs of the zodiac, with the net of heaven stretched between them. Likewise, Edge's work on the Hall of Bulls at Lascaux and the presence of many of the elements of astronomical myth among New World peoples argue for a much earlier date. So do the Great Sphinx of Giza and the three pyramids.

Return to Egypt

Why?—this is the question that hangs still over the Sphinx. Why did the ancient Egyptians build these massive monuments? The level of effort seems almost unimaginable, particularly for a people who, as far as we know, had nothing but tools of copper, bronze, stone, and wood. The Great Sphinx is the largest monumental sculpture known from such an early date, yet it is dwarfed by the later pyramids. The largest of the three, the one attributed to Khufu (also known as Cheops), is mostly solid masonry comprising two and a half million limestone blocks. At an estimated average weight per block of 2.6 tons, the total weight exceeds 6 million tons. The pyramid of Khafre (Chephren) is smaller, about 5.25 million tons, and the one attributed to Menkaure (Mycerinus) is by far the smallest of the three, representing only an approximate 600,000 tons. And there is more to Giza than the three pyramids and the Great Sphinx. The site contains a variety of causeways, temples, miniature pyramids, boat burial pits, tombs, and other structures.

Giza represents a commitment of extraordinary resources of time, energy, human labor, engineering skill, and artistic accomplishment that becomes even more remarkable when compared with more modern religious monuments. For example, Khufu's pyramid is only twenty-four feet higher than the dome of Saint Peter's in the Vatican, yet it covers thirteen acres, compared to Saint Peter's mere four. Were Khufu's pyramid hollow, it could easily enclose Saint Peter's, with enough space left over that a rearranged Westminster Abbey would fit inside as well.

The Egyptians were fascinated with death and the possibility of life beyond the mortal moment, and this fascination is considered central to Giza and the other ancient Egyptian temples and holy sites. As a result, the pyramids have long been considered outsized tombs constructed to give the pharaohs buried within them their very best chance at immortality.

Various scholars have wondered, however, whether there isn't more to the pyramids than elaborate tombstones. In itself, this idea is hardly revolutionary. The great medieval cathedrals of Europe were the burial places of many a king, earl, duke, baron, bishop, and abbot, yet these immense and impressive buildings can hardly be dismissed as overdone graveyards for secular and ecclesiastical elites. In the case of the cathedrals, we know better what purposes these buildings served because the culture of medieval Europe remains comprehensible to us. Egypt five millennia in the past, though, is another matter altogether. The very alienness of that distant and lost world makes it difficult for us to understand wholly what the Egyptians were up to. If the pyramids served a purpose other than burial marker, what was it?

Some interesting ideas on these issues have been offered by Robert Bauval and Adrian Gilbert in *The Orion Mystery,* and by Graham Hancock and Robert Bauval in *The Message of the Sphinx.* In many ways, these books leave something to be desired. None of the authors is a scientist, and all three are far too eager to inflate problems into "mysteries" and puff interesting insights into revolutionary findings that are nothing of the sort. In the end, too, they turn the Sphinx into a kind of astronomical scavenger hunt, purportedly pointing to the existence of a treasure trove of ancient records in a secret cavern hidden deep under the Sphinx. In the course of our seismic work, Thomas Dobecki and I did find some kind of void or chamber in the deep rock under the Sphinx's left paw, but I find it ridiculous to conclude in advance that this area, which may be a natural geological feature and not the result of human effort, contains a legacy of wisdom passed down from some long-lost source, perhaps even Atlantis. Still, these three authors have developed ideas that, while far less definitive than they claim, may point to an earlier date of origin for the Giza Necropolis than the orthodox view allows.

This same orthodoxy has held that Egypt's religion centered on the sun, particularly as manifested in the god Osiris, consort of the Earth and moon goddess Isis. Bauval, Hancock, and Gilbert advance a contrary argument. They maintain that in the ancient Egyptian cosmology the night sky was the truly important element and that, as in the monomyth investigated by Santillana and Dechend, the stars above determined the fate of the world below. The curious ground plan of Giza, with its seemingly misaligned pyramids, is a hologram for the three stars forming the belt of the constellation Orion, which was associated with Osiris. Even the Nile played a role, standing for the Milky Way, which was thought in ancient times, according to Santillana and Dechend, to be the resting place for souls awaiting their next incarnation. Additionally, Giza turns into a kind of giant clockwork marking the night sky of 10,500 B.C., an era when the star of Isis—the one we know as Sirius, the Dog Star in Orion—would rest precisely on the horizon. The causeways running from the pyramids, according to Hancock and Bauval, mark the precise rising point of the sun on the cross-quarter days between the solstice and the equinox, in both summer and winter. As for the Sphinx, its pitilessly blank gaze fixes eastward on the rising point of the constellation Leo—fittingly, the Lion, the Sphinx's own body shape—on the winter-spring cross-quarter.

The precision to which Hancock, Bauval, and Gilbert lay claim is indefensible. Computer simulations of the sky at the remove of several millennia, connected to survey-type readings taken from very old, very worn monuments, are markedly less exact than the three authors would have us believe. Still, they are right in pointing up the powerful astronomical associations of Giza, which make even more sense when considered with the new finds at Nabta. In southern Egypt, thousands of years before the beginning of pharaonic Egypt in circa 3000 B.C., there lived a people fascinated by the heavens, given to careful observation of the movement of the sun, adept at moving large slabs of stone, and committed to the construction of monumental structures—all the attributes of Giza. It might well be that these vanished people are in fact among the ancestors of the ancient Egyptians, and that they brought their finely tuned astronomical sense to the cultural and religious mix that gave birth first to the Sphinx and later to the pyramids.

Hancock, Bauval, and Gilbert are right, too, in showing how the sky references at Giza point to a time well before the Old Kingdom building dates of the Giza pyramids, as does the Great Sphinx itself. This hypothesis does not mean that the pyramids as we know them were built at this time, however. It is possible that the pyramids were attempts to fix an original, unsullied sky—the heavens of the First Times, as Hancock, Bauval, and Gilbert call them—a kind of celestial Garden of Eden that existed before the slow decay of precession set in. Temples and holy buildings, like Chartres cathedral and the Taj Mahal, commonly serve the purpose of providing an earthly experience of the ultimate reality of eternal paradise. The pyramids are likely no different. And it is also possible that they were constructed to incorporate earlier monuments of an age similar to the Sphinx's as a way of capturing their sacred energy, just as many of the cathedrals and shrines of Europe, Chartres and Kildare among them, were built atop older, pre-Christian holy places.

A New Look at the Old Stories

None of the evidence we have examined is absolutely definitive, yet the trend of information is undeniable and convincing. The redating of the Sphinx, the ancient astronomical alignments of Nabta, the sophisticated ancient cities of Jericho and Çatal Hüyük, the tripling of the duration of humankind's presence in the Americas, the star maps of Lascaux, and the knowledge of precession indicated in world myth and in the Giza pyramids all point to a much earlier date for civilization than we have believed possible.

This understanding raises a new line of inquiry. Many societies pass on myths—there's that significant word again—about remarkable cities of long ago, legendary places of fabled wisdom and wealth that flourished in a distant bygone time and vanished. Is there more to these tales than exotic entertainment? Are they in fact the storied remnants of earlier civilizations, ones that disappeared, perhaps in catastrophe, and took with them ancient ways of being we should know more about?

Looking for the Lost Cities

OUR INVESTIGATION INTO THE EVIDENCE FOR A LOST civilization begins not with physical evidence like that presented by the Great Sphinx of Giza, but with two of the most powerful minds in the history of humankind—the ancient Athenians Socrates and Plato. The modern search for the lost continent of Atlantis opens in the surviving literary record of an extended philosophical conversation conducted by Socrates and recorded by Plato over two millennia in the past.

Atlantis: The Original Lost Continent

As a gauge of interest in Atlantis, the Czech geologist Zdeněk Kukal undertook the formidable task of tabulating the mass of publications, both academic and popular, covering the topic. By Kukal's reckoning, in the years between 1890 and 1980, almost 43,800 book and periodical pages—enough to fill approximately 175 volumes the size of this one—dealt with finding the lost continent. Interest in the topic has hardly flagged in the intervening years between Kukal's count and today. As I began writing this chapter, I did a search of the Internet and came up with over 200,000 World Wide Web pages that drew on the name Atlantis and the ongoing interest in this mystery.

Remarkably, this prodigious commentary is based on a very small—and incomplete—literary record. Plato was the first to mention Atlantis, specifically in two of his dialogues, both written circa 360 B.C.: the *Critias,* which he never finished, and the *Timaeus,* which deals with Atlantis almost in passing and devotes more of its energy to a consideration of the ideal political state and the nature of the universe. For a writer who was both prolific and loquacious, Plato didn't really have all that much to say about Atlantis, yet his small literary foray into this story has given rise to a long, continuing, even obsessive interest in identifying and locating the lost continent.

This interest is primarily a modern phenomenon. The literature of the ancient world includes various other mentions of Atlantis, but all of them derive directly or indirectly from Plato. For example, Diodorus Siculus, a Roman historian of the first century B.C. who lived in Sicily, relates in his massive forty-volume *Historia Bibliotheca* that the Amazons defeated the Atlantians in war, then joined with them to conquer the tribe of the Gorgons. Some writers have taken Diodorus's putative account as an independent corroboration of Plato. It's nothing of the kind. Diodorus lifted the story from the Greek historian Herodotus (484?–425? B.C.), and it concerns not Atlantis, the lost island continent, but the warlike doings of tribes who lived at the foot of North Africa's Atlas mountain range. As for Diodorus's description of Atlantis as a beneficent and wondrous island located in the great sea beyond the Pillars of Heracles, it is clearly copied from Plato's earlier account. In all fairness to Diodorus, he was not a plagiarist. Borrowing from earlier stories was a common and accepted practice in the ancient world, a technique Plato himself, like other classical writers, used when it suited his philosophical purposes.

The medieval period demonstrated little interest in Atlantis. The Middle Ages centered more on religion than mythology, and Plato was hardly read in an age dominated by Aquinas's version of Aristotle's very different way of thinking.

Atlantis makes one of its few appearances in the period before the nineteenth century on a map by the German Jesuit Athanasius Kircher that dates to 1665. Kircher placed the continent in the North Atlantic, between Europe and North America, and labeled the area as "the site of Atlantis, now beneath the sea, according to the beliefs of the Egyptians and the description of Plato."

It is only in modern times, and particularly our own century, that interest in Atlantis has blossomed fully. Unfortunately, the whole topic has become a mélange of fantasy as well as fact, with various sorts of fringe New Agers, occultists, true believers, and pseudoscientists seizing upon Atlantis as the lost repository of a wisdom our ailing world desperately needs. The need for such wisdom is self-evident. By contrast, the existence of Atlantis is a fact that can be proved or disproved. Determining the truth of the matter may be the first step toward discovering the wisdom we desperately need.

Plato on Atlantis

Originally educated as a dramatist, Plato (427–348 or 347 B.C.) left the theater to become a disciple of Socrates (469–399 B.C.). This was a radical, even dangerous move, since Socrates was later condemned to death for corrupting the morals of the youth of Athens. Plato remained faithful to his mentor, however, recording his memory of the teachings of Socrates, who, like Jesus and Gautama Buddha, left no written record. In this effort, Plato made profitable use of his background in drama, writing not in the form of a philosophical treatise, but in a fictionalized dialogue. Plato portrays Socrates as the greatest seminar leader in the history of humankind, a wise and able interlocutor who leads his students to examine their ideas and then demonstrates to them the accuracy or error of their thinking. Plato's two accounts of Atlantis occur in this literary setting.

Both the *Timaeus* and the *Critias* concern the same cast of characters. The central figure is, of course, Socrates. Then there is Timaeus, of whom no historical record exists and who may have been Plato's invention; Critias, who was Plato's great-grandfather; and finally Hermocrates, a statesman and soldier from the Greek colony of Syracuse in Sicily, a city where Plato spent an important period of his life.

The unfinished *Critias* is largely a recounting by Critias of an ancient story remembered from his youth. Memory is a key issue in the telling of this tale. Advised by Hermocrates to invoke the good offices of Apollo and the Muses before he begins to speak and "show forth the virtues of your [Athenian] citizens," Critias asks as well for the favor of Mnemosyne, the goddess of memory, to ensure that "I can recollect and recite enough of what was said by the [Egyptian] priests and brought hither by Solon." Solon (638–559 B.C.) was a lawgiver, poet, and traveler, something of an Athenian Thomas Jefferson. By tracing the origin of the tale to Solon, Plato gives it great prestige and authority.

Critias begins by saying that according to the story told by Solon, 9,000 years had passed since the great war pitting all those who lived outside the Pillars of Heracles against those who lived inside it. The kings of Atlantis commanded the outsiders; the insiders came under Athenian leadership. In the end, a great earthquake sank the island nation of Atlantis, which was larger than North Africa (which Critias, like all Greeks, called Libya) and the Middle East (which to him was Asia), and became a barrier to navigation.

Next, Critias tells about the inhabitants of ancient Athens, where "military pursuits were then common to both men and women" and which was ruled over by the god Hephaestus and the goddess Athena, who gave the city her name. In those days the land was richer, deep in soil, the mountains fully forested. All that changed in the many deluges of the period, which were so severe that in but one night the original Athenian acropolis was washed away, as was much of the topsoil and many of the forests.

Atlantis fell within the portion of the world governed by Poseidon, the trident-brandishing god of the sea and of horses. Atlantis, like Athens, was a country well blessed. The center of the mountainous island continent was taken up by a large and fertile plain, which Poseidon turned into a paradise after he fell in love with a mortal woman named Cleito and fathered the Atlantian race through her. Poseidon ringed the hill where Cleito dwelt with alternating circles of sea and land so that the island and his mortal lover were safe from invasion. In the plain he raised two springs of water, one hot and one cold, and he blessed the soil with every kind of produce, so that the Atlantians had food in both variety and abundance. As they multiplied through the years, the children of Poseidon and Cleito became a royal race "who were the inhabitants and rulers of divers islands in the open seas; and also... they held sway in our direction over the country within the Pillars as far as Egypt and Tyrrhenia [a region of Italy]." The Atlantians were unspeakably rich; "they had such an amount of wealth as was never before possessed by kings and potentates." They used their great riches and the natural abundance of the island, which supported every sort of animal, including elephants, to build a great metropolis well furnished with temples and shrines, roads and canals, fountains and pillars, citadels and guardhouses. This was a religious people, whose liturgy centered in part on wild bulls that were sacrificed within the temple of Poseidon as a pledge of faithfulness to the ancient laws written in stone by the god.

In time, though, this idyll came to an end. Slowly the divine aspect of the Atlantians faded away as Poseidon's portion became more and more diluted in each successive generation. Invariably the Atlantians slipped off the path of virtue. According to Critias, "[W]hen... human nature got the upper hand, they then, being unable to bear their fortune, behaved unseemly,... and they appeared glorious and blessed at the very time when they were full of avarice and unrighteous power." Zeus resolved to punish these degraded people for their sins, and he called the gods and goddesses into council to select an appropriate comeuppance.

And there, abruptly, the *Critias* ends.

The *Timaeus* opens with Socrates reminding the participants in the dialogue of the topic of conversation begun the day before: the perfect state. Timaeus and Socrates review the various attributes of a well-run polity, creating a persuasively utopian image of justice, rectitude, and virtue. At the invitation of the others, Critias tells Socrates the story he has earlier shared with Timaeus and Hermocrates, an ancient tale that Solon first heard from Egyptian priests in the holy city of Sais and later passed on to Critias's grandfather, who told it to him. This tale, Critias maintains, concerns "the greatest action the Athenians ever did, and which ought to have been the most famous, but, through the lapse of time and the destruction of the actors, it has not come down to us." In Plato's time, as in our own, Atlantis was a lost story as well as a lost land.

According to what Critias heard indirectly from Solon, the Athenians of Plato's time were but a remnant of the city's original inhabitants, the few who survived the many terrible deluges that marked the period of the war with the Atlantians. The Athenians did not go to war for reasons of greed or expansion, as the Atlantians did. They took up arms to meet the terrible threat of the Atlantians, who invaded Europe and Asia without provocation. Based on their island outside the Pillars of Heracles, which again Plato makes the size of North Africa and the Middle East combined, the Atlantians obtained control of the Mediterranean as far as Egypt and Tyrrhenia and prepared to deal a death-blow to the Athenians, who stood alone against the threat, abandoned by erstwhile allies. Then came violent earthquakes and floods, which destroyed all of Athens's fighting men by sucking them into the earth and, both simultaneously and fortuitously, sank the island of Atlantis. All this took but "a single day and night of misfortune."

Following his description, which occupies only a fraction of the complete dialogue, Critias passes the baton to Timaeus, who opens a discussion on the nature of creation. Atlantis is not the point of the *Timaeus.* It is an example Plato cites to draft his philosophical argument, an important point to which we shall return.

The Search for the Lost Continent

In the *Critias,* Plato has Socrates underscore the truth of the tradition related by Critias, who asks the philosopher whether the story suits the purpose of the group's philosophical soirée. Socrates makes clear that it does: "And what other [story], Critias, can we find that will be better than this, which...has the very great advantage of being a fact and not a fiction?" The many searchers for Atlantis have taken Socrates—or, more accurately, Plato—at his word, basing their explorations and hypotheses on the assumption that Atlantis is indeed historical and not fictional.

So, where was Atlantis? And what terrible and sudden catastrophe caused it to sink beneath the sea and there be lost until Solon heard about it from the Egyptian priests of Sais, 9,000 years later?

The hunt for a real, true, factual Atlantis has historical precedent in the search for both Troy and Minoan Knossos. In the nineteenth century, practically all classicists and ancient historians, who practiced their disciplines well before the development of modern archaeological methods, accepted as fact the concept that Homer had invented the ten-year siege of Troy, whose final battles are told in the *Iliad.* Still, Troy-hunting was something of a pastime among certain Europeans with the right combination of leisure, money, and classical training. One of them, an amateur student of ancient history named Heinrich Schliemann, identified and excavated the mound along the Turkish coast on which the citadel of Homer's Troy had stood. Announcing his discovery to the world in the late 1860s, Schliemann went hunting for the other buried treasure of antiquity, Knossos. This Cretan palace, built by the Minoans, boasted the famous Labyrinth that enclosed a monstrous half-man, half-bull known as the Minotaur, to which the Minoans required the sacrifice of fourteen Athenian youths and maidens every nine years. The Labyrinth was said to have been designed under duress by the Athenian Daedalus, who, homesick for his native land, fashioned wings from wax and feathers for himself and his son, Icarus, to fly back to Athens. Icarus flew too close to the sun, which melted his wings and dropped him into the sea as Daedalus watched helplessly. Later, the great Athenian hero Theseus ended Minoan domination of Athens by piercing the maze of the Labyrinth, slaying the Minotaur, and, with the aid of the Minoan king's daughter, making his way back out. Schliemann came very close to finding Knossos, locating a likely site near the city of Heraklion, but his death put an end to the search. Working independently, Arthur Evans unearthed Knossos in the 1890s, earning himself British knighthood for his achievement.

Is Atlantis a real location like second-millennium B.C. Troy and Knossos, as Plato himself seems to assert, a factual locality of old awaiting modern discovery? Or is it a philosophical fiction?

Unfortunately, Plato offers the only literary record of Atlantis. Although his description of the Atlantian city and its religious practices is explicit and detailed, his geography leaves a great deal to be desired. The account is at best ambiguous, at worst contradictory. He speaks of a huge island situated outside the Pillars of Heracles and boasting a climate tropical enough to support elephants. Beyond Atlantis lay another, larger body of land that Plato calls the "opposite continent." Locating Atlantis would seem to be a simple issue of sailing through the Pillars of Heracles and looking for the remains of a sunken tropical island with a major landmass off its opposite shore. There arises the first of many problems: Where are the Pillars of Heracles?

This distinctive name, taken from the most powerful hero of Greek mythology, was given to a number of ancient sites known in modern times by quite different appellations. One of the most common is the Strait of Gibraltar, which separates the Mediterranean Sea from the Atlantic Ocean. If Plato meant Gibraltar, Atlantis would lie somewhere around the Azores, the Canaries, or Madeira, with North America filling the role of the opposite continent. The Greeks, however, used the name Pillars of Heracles to mark other sites besides Gibraltar, some outside the Mediterranean—namely, the Canary Islands in the Atlantic and the Strait of Kerch dividing the Black Sea from the Sea of Azov—and even more inside—specifically, the Strait of Bonifaccio between Corsica and Sardinia, the Strait of Messina between mainland Italy and Sicily, the Greek Peloponnese, the mountainous coast of Tunisia, and the Nile Delta. Obviously, moving the Pillars of Heracles inside the Mediterranean poses an enormous problem, since the opposite continent could be either Africa or Europe. Likewise, the name Atlas, the root from which Atlantis is derived, was identified with a number of mountain ranges, including the Atlas range of Morocco, the northern mountains of the Arabian Peninsula, the Caucasus, and the Nubian highlands.

As a result of all this confusion over names, a long list of mutually contradictory sites for Atlantis has been proposed in the voluminous literature on the topic. There are at least nineteen possible locations in Europe and North Africa: the southwestern Spanish coast (where the ancient Phoenicians had a colony); the Atlas Mountains; the central Sahara; the Lake of the Tritons in Tunisia; Corsica; Sardinia; the Tyrrhenian Sea; Sicily; the Ionian Sea; the Aegean Sea; Thera (Santorini); Crete; Rhodes; the Azov Sea; the Black Sea; the Levant (Israel and Lebanon); Egypt; Troy (northwestern Turkey); and the North Sea coast of Holland and Belgium. Some Atlantis enthusiasts are convinced that the lost continent lies far outside the ancient world of Europe and North Africa. They have proposed a long list of more distant candidate locations, including Greenland, Iceland, Scandinavia, eastern North America, Bermuda, the Azores, the Canaries, the Cape Verde Islands, the Mid-Atlantic Ridge, the Yucatán Peninsula, the

Caribbean Sea, the Atlantic and Pacific coasts of South America, and the ancient kingdoms of Zimbabwe in southern Africa or Ghana in west Africa. Atlantis has also been placed at the confluence of the Tigris and Euphrates rivers, which some people hold to be the site of the biblical Garden of Eden. The Indonesian archipelago is currently being advanced as a candidate, and, as we shall explore further, Antarctica has become a popular possibility, one backed by the work of a serious scholar.

Given the wide range of nominated localities, covering both hemispheres and stretching nearly from pole to pole, there is more than a little confusion over Plato's ambiguous directions. And, when we look carefully at each candidate, we find they all come up lacking in some major way. A few examples bear examination.

Ignatius Donnelly, who wrote the Atlantological "bible" of the nineteenth century, proposed that the continent had sunk under the swells of the Atlantic about midway between Europe and North America. This possibility does seem to fit with Plato's description of a site outside the Pillars of Heracles, assuming the philosopher meant Gibraltar, and between Europe and the opposite continent, assuming he meant North America. The problem, though, is that the floor of the North Atlantic has been mapped in the century since Donnelly wrote, and it is very clear that there is no submerged, continent-sized landmass anywhere in all that world of water. Indeed, the Mid-Atlantic Ridge is actually rising, not slipping deeper into the sea.

A number of scholars, following the lead of the Greek archaeologist Spyridon Marinatos, have advanced a creative and interesting argument that Atlantis is in fact Minoan culture, and that the devastating sinking of the island in a day and a night refers to the explosion of the volcano on the island of Thera, not quite seventy miles to the north of Crete. Again, certain details fit. The Minoans, like the Atlantians, worshiped bulls. The walls of Knossos were decorated with magnificent frescoes showing athletic young men and women flinging their lithe bodies over the horns of bulls large enough to make the bravest matador shiver in fear. Like Atlantis, Knossos was extremely wealthy. A sea-trading people, the Minoans had the resources to lavish themselves and their cities with the best of everything, building a great palace that in some ways resembles the description of the royal abode in the *Critias*. And Thera—which is also known as Santorini, a corruption of Santa Irene, the name given to it by Italian sailors when the island belonged to the medieval Venetian empire—was actively volcanic in ancient times. Thera itself, along with four smaller islands in its bay, once constituted a large single volcano that collapsed into a massive caldera, or volcanic crater. A cliff rising 1,100 sheer feet from the Thera harbor to the village above serves as a dramatic reminder of the size and power of the volcano that once hurled its smoke into the sky.

One Atlantian hypothesis is that Thera detonated in a single massive explosion, which sent a monstrous tidal wave, or tsunami, hurtling toward Minoan Crete. Since the Thera caldera is about three times the size of the crater left by the volcanic explosion of Krakatoa—which occurred in 1883 and whose effects are well documented—the assumption is that a tidal wave three times the size of Krakatoa's struck Crete. That means a wave more than 300 feet high, which would have hit Crete within half an hour and rolled into Egypt, the Levant, and the Syrian coastline three hours later, all with devastating consequences. Knossos and the Minoans would have been swept away in seconds, satisfying Plato's Atlantian criterion of quick destruction.

Several notable problems arise with the Knossos-as-Atlantis hypothesis, however. The first is time. Plato is clear that the war between Athens and Atlantis occurred 9,000 years before the Egyptian priests passed the story on to Solon, which works out to circa 9600 B.C. Some Atlantis enthusiasts have proposed that Plato made a decimal error, and that he meant to write 900 instead of 9,000. If so, his dating, which would be 1500 B.C., would fall fairly close to the radiocarbon date of 1410 B.C. ± 100 given to the Thera eruption.

I find this explanation fanciful, however. The Greeks of Plato's time used a clumsy system of arithmetic notation that lacked a symbol for zero. Plato couldn't simply drop a space and inadvertently turn 9,000 into 900.

Also, like other well-educated ancient Greeks, Plato was fully aware of the Minoans. The civilization of the Greek mainland drew heavily from Crete, and vice versa, as shown in the myth of Daedalus's involvement in designing the Labyrinth and the hero story of Theseus. If Plato wanted to write about Knossos, why didn't he call it Knossos instead of inventing the name Atlantis? Why would he make Atlantis so much larger than Crete? And where did the elephants come from?

Finally, geological evidence thoroughly weakens the tsunami hypothesis. A detailed study of the volcanic ash deposits on Thera indicates that the volcano didn't detonate in one massive explosion like Krakatoa or Mount St. Helens. It collapsed in slow stages that threw huge amounts of sun-dimming ash into the air but produced no detonation big enough to yield a 300-foot tidal wave. The city of Akrotiri on Thera physically survived the explosion, preserved, like Pompeii, under a thick deposit of ash, yet lacking human remains. How could it be that Akrotiri stood and its inhabitants apparently escaped the devastation, while Knossos drowned? In fact, the Minoan civilization, though in decline, did continue for some time after the Thera eruption, until it was eventually absorbed by the mainland Greeks. Even though the Thera eruption may have made life on Crete hard—causing earthquakes and torrential rains, as volcanic events often do—it by no means marked the end of Knossos.

All in all, Knossos doesn't qualify any better than the Mid-Atlantic Ridge as an Atlantis candidate.

The Twilight Zone

One of the major difficulties in unraveling the Atlantis mystery is separating fact from fiction. Atlantis has attracted tremendous interest from those devoted to occult practices, an unwavering belief in unidentified flying objects, the psychic readings of Edgar Cayce, and similar faiths. My coauthor told me of the chiropractor he once interviewed for a magazine article who told him that the separation of mind and body, which wreaks such havoc on human health, owed its origin to the dictatorial psychological-control methods of the Atlantian ruling class.

"And how did you learn this?" Robert asked him.

"It was channeled from the Atlantian elders," he said. "This isn't the sort of thing you find written down anywhere."

No, it isn't. And unfortunately this kind of sloppy thinking is all too common in much of the writing on Atlantis. An example I encountered recently comes from the theories proposed by Arysio Nunes dos Santos, which places Atlantis in the South China Sea among the many islands of Indonesia. Not only is this the location of Atlantis, according to Santos, but it is also the site of "the Garden of Eden, the Island of Avalon, the Garden of the Hesperides, the hideout of the New Jerusalem, the true location of Troy and of Lanka, as well as the Holy Land and Paradise that has been promised us all from the dawn of time"—a heavy load of mythological weight to be borne by a sunken island, even one the size of a continent. Santos develops his argument with a series of remarkable identifications. Mount Atlas, the main mountain of Atlantis (Plato never names it) is Krakatoa, whose massive explosion—apparently 1883 was a repeat of an earlier, primordial event—opened the Sunda Strait between Java and Sumatra. All this makes sense, Santos says, when you realize that the Greeks originated in the Indies and migrated into the Balkans, where they superimposed their old geography on a new homeland.

Santos does get one thing right: At the time Plato gives for Atlantis, much of the South China Sea would have been dry land, covered over by a major rise in sea level at the end of the most recent ice age—an event that is critical to what I find to be the best working explanation of Atlantis. The rest of it just doesn't work, not even remotely. In truth, there is no evidence that the Greeks came from Southeast Asia, an area whose language groupings are utterly different from the Hellenic tongue, nor did they venture so far from home. When Alexander the Great took his army into India, he wanted to cross the subcontinent to see the ocean he, like other educated Greeks, thought was the edge of the world. Indonesia lies, of course, beyond this boundary, and Greek ignorance of this substantial part of the globe speaks volumes about their origins—and the utter unlikelihood of Atlantis among the islands of Indonesia.

An even more striking example of fantastical, paradisiacal thinking about the lost continent comes from the curious history of the other lost continent—Lemuria, also known as Mu. Promoted by people like twentieth-century authors James Churchward and Javier Cabrera, Lemuria is said to be an Atlantian-sized area of some 15,000 square miles and a population of 64 million people that sank into the Pacific Ocean over 50,000 years ago. Today all that remains of Lemuria are its mountain peaks, which poke up out of the sea to form such islands as Hawaii, Fiji, and Tahiti. According to the legend, Lemuria was a tropical paradise, a South Pacific Garden of Eden whose inhabitants lived free of conflict and disease in homes with transparent roofs, apparently unafraid of baring all to their neighbors. Adept at extrasensory perception and mental telepathy, the Lemurians were vegetarian, agricultural, and completely organic, living in harmony with nature.

According to an account written by the Rosicrucian historian W. S. Cerve, the Lemurians subjected couples who wished to marry to a demanding compatibility test. The man and the woman had to surrender all their possessions, including clothing, to a priest and retreat naked into the forest, there to live away from human habitation for twenty-eight days. During this time alone the couple had to build their own shelter, fashion clothes and tools, and forage for food, all without engaging in conflict or enmeshing each other in negative thinking. Only if the couple could succeed at this daunting task were they allowed to marry. Otherwise it was back to singlehood, ideally with some new wisdom gained from the wilderness experience.

The legend holds that when Lemuria sank into the Pacific, not all of its great wisdom was lost. Maps of the continent were preserved in pre-Inca Peru, where they were discovered by Javier Cabrera. The Lemurian elders and their wisdom remain available for consultation through psychic channeling. And the survivors of Lemuria fanned out over the face of the Earth to become the Tibetans, the Mayans, the Native Americans, and the Eskimos, peoples who keep alive oral traditions of great antiquity.

Yet there's nothing truly old about the stories of Lemuria. This lost continent is a modern invention, the remains of an abandoned nineteenth-century scientific hypothesis turned into a popular and mistaken belief.

In the 1860s and 1870s, a group of British biologists noticed unexplainable geological similarities in India, southern Africa, South America, and Australia. In these widely separated areas, certain strata of the Permian age, dating back to 245–286 million years ago, contained almost exactly the same kinds of sedimentary rocks and included identical fossils of land plants and land animals. In the late nineteenth century, long before the theory of continental movement by means of plate tectonics (an idea we will explore in chapter 5), geologists assumed the continents were immobile and fixed in place. This left a quandary: How could land animals and plants have crossed such immense stretches of open sea at some point in the distant past? The answer: Land bridges once connected the contemporary continents, provided the plants and animals with passage, then later sank into the sea.

Ernst Heinrich Haeckel, a strong advocate of Darwin's ideas, used the land-bridge concept to explain the presence of lemurs—distinct, early forms of primates—in Africa, India,

Madagascar, and Malaysia. This particular land bridge, he proposed, had remained above water long enough for these animals to have spread from their point of origin into regions now widely separated. Philip L. Scalter, an English biologist, picked up on this association with lemurs to dub the land bridge Lemuria.

As a scientific concept, Lemuria lost credibility when scientists came to the modern understanding of continental movement by plate tectonics and its effects on the distribution of plant and animal species. Occultists picked up the concept, however, and used it to their own ends. The most successful of these was Madame Blavatsky, a prominent occultist and theosophist of the late nineteenth century. Her *Secret Doctrine* (first published in 1888), which mixes science and occult knowledge with the Rig Veda and other Hindu sources, made Lemuria into a lost continent in the Indian Ocean populated by apelike creatures that laid eggs and contained both sexes in their bodies. Later writers like Annie Besant (*Man's Life in This and Other Worlds* and *The Pedigree of Man,* among other titles) and W. Scott-Elliot *(Legends of Atlantis and Lost Lemuria)* embellished on Blavatsky. In these new tales of lost continents, Lemuria was home to dinosaurs and bronze humanoids up to fifteen feet tall. Now, in its most recent incarnation, Lemuria has moved from the Indian to the Pacific Ocean.

Both Atlantis and Lemuria are prime examples of the workings of what psychoanalysts call projection. In projection, an individual takes some portion of his or her psychic self and attaches it to an outside individual or institution. Projection is the process that allows us to make enemies in warfare evil, and to paint them as deserving of the violent death we are about to deliver. Likewise, projection can be used to create a utopian picture of what we most desire, such as a paradise built on wisdom, innocence, and childlike trust. It strikes me as hardly accidental that Cerve's account of the marriage customs of the Lemurians is much like the premarital practices Thomas More invented in *Utopia.* The point isn't that Cerve was copying; it is that his account of

Lemuria and other such stories constitute a utopian exercise, not history.

Antarctic Atlantis?

Amid all the fantasy, projection, and inaccurate fiction that has attached to the topic of Atlantis, one serious idea deserves a rigorous look. This is the concept that Atlantis was located in the Antarctic. The theory originated in the scholarly work of Charles Hapgood, which forms the foundation of the contemporary popular book *When the Sky Fell: In Search of Atlantis,* by Rand and Rose Flem-Ath. Antarctic Atlantis also appears as the concluding idea of Graham Hancock's *Fingerprints of the Gods: The Evidence of Earth's Lost Civilization,* and it figured prominently in the television special "The Mysterious Origins of Man," narrated by Charlton Heston and featuring Hancock's ideas and on-camera presence.

Hancock, Hapgood, and the Flem-Aths base their theory on the interpretation of a number of maps made in the late medieval period. What these maps show—or do not show—is key to evaluating the hypothesis that Antarctica is the site of a civilization now lost to us, the one that Plato called Atlantis.

The most striking, and the best known, of the charts is the Piri Reis map, which was discovered accidentally in 1929 by scholars working in archives of the Ottoman empire stored in the Topkapi museum in Constantinople. Dated to 1513 and signed by the Turkish captain Piri Reis, the map claims to contain information about the New World derived from the voyages of Christopher Columbus. An inscription on the map, drawn on a gazelle-hide parchment, says, "The coasts and islands on this map are taken from Colombo's map." The discovery of the map, with information so close in date to Columbus's discovery of the New World, caused a scholarly stir when it was announced.

The story of the map began in 1501, just nine years after Columbus's first voyage to America, when Kemal Reis, another

Ottoman captain and Piri Reis's uncle, captured seven ships off the Spanish coast. In those days, navigational information about the newly discovered lands of the Indies and the Americas was extremely valuable and carefully guarded. Kemal Reis interrogated the Spanish crewmen in the hope of uncovering trade secrets, and he was pleased to discover one sailor who had accompanied Columbus. Even better, the sailor had in his possession a map said to have been drawn by Columbus. Reis seized it and willed it to his nephew.

In 1511 the younger Reis set himself to the formidable task of drawing a map of the world that included all the Spanish and Portuguese discoveries and that was based on some twenty sources, one of which was the map of the New World captured by Uncle Kemal. Reis also used a number of maps purportedly dating to the fourth century B.C.—the time of Alexander the Great—an Arab map of India, and Portuguese maps of the Indian Ocean and China. Reducing all the maps to a single scale, which in those days was a difficult challenge, Reis spent three years assembling his chart, and presented it as a gift to the Sultan Selim.

This proved to be a wise career move, at least in the short term. Reis was promoted from captain to admiral, and he wrote a book that provided a mariner's guide to the islands and coasts of the Mediterranean and advocated driving the Portuguese out of the Red Sea and the Persian Gulf, where they threatened the Ottomans. Reis got his chance to do just that in 1551, but he lost most of his fleet in battle. Though an old man in his eighties, the disgraced Reis was beheaded at the order of an unforgiving sultan.

After the initial stir of its discovery, the Piri Reis map lay largely uninvestigated until Charles Hapgood, a Harvard-educated science historian teaching at New Hampshire's Keene State College, gave his class the assignment of studying it in detail. That assignment resulted ultimately in Hapgood's classic book *The Maps of the Ancient Sea Kings.*

Hapgood came to the Piri Reis map not because he was interested in ancient history, but because he had come to believe in a theory holding that Earth's crust can slip rapidly over the underlying mantle, causing sudden shifts of geographical location and a complete, even catastrophic, change of climate—an idea that will be examined in detail in chapter 5. Hapgood was working on his crustal-shift theory when he heard of a curious claim about the Piri Reis map made by Captain Arlington H. Mallery, a U.S. Navy cartographer, engineer, and ancient-map specialist. Mallery asserted that the old chart showed the coastline of Antarctica and, even more dramatically, it depicted that coast as free of ice cap and glacier.

Mallery's idea very much intrigued Hapgood. Antarctica wasn't discovered until 1820, 307 years after Piri Reis completed his map. How could the Turkish captain have included a landmass no one knew existed? And why had he shown it without ice? For that matter, how could he have known how it looked with the thick glaciers removed? And if the map did depict Antarctica, why did Piri Reis write on it, "There is no trace of cultivation in this country. Everything is desolate, and big snakes are said to be there. For this reason, the Portuguese did not land on these shores, which are said to be very hot"? Big snakes? Hot? Antarctica?

Hapgood put his students to work on these ideas, and they turned up several puzzles. For one, the map seemed to show the Andes along the western portion of what is obviously South America, complete with a creature that had a passing resemblance to a llama, an animal found only in the Andes. Again, timing didn't work. Magellan found a way around South America in 1520, seven years after Piri Reis's map, and Pizarro first sighted the Andes in 1527, fourteen years after. Hapgood developed an explanation: One of Reis's source maps, probably one of the charts dating to the time of Alexander the Great, showed the Andes. If this idea was true—and for Hapgood, the presence of the putative llama clinched it—then someone had to have known about the Americas at least 1,800 years before Columbus.

Hapgood's students also identified what they said were the Falkland Islands—undiscovered until 1592. And if those islands were the Falklands, then the land to the south had to be the coast of the Queen Maud Land region of Antarctica. Pursuing this line of thought, Hapgood and his students concluded that the map did indeed bear a striking resemblance to Queen Maud Land, and that the chart was accurate within twenty miles.

Hapgood remained cautious, however. He was well aware that in ancient and medieval times, many mariners and cartographers believed in a large landmass lying in the distant, forbidding reaches toward the South Pole. It was possible, therefore, that Piri Reis was simply following this tradition and engaging in wishful geographical thinking. And the map did have one big problem: It showed South America and Antarctica as a continuous coastline, as if the two continents abutted. In fact, there are over 600 miles of open ocean between their closest points.

Working in the Library of Congress in 1959, Hapgood came across another old map that wiped out all his doubts and confirmed his ideas. The chart, drawn in 1531 by the French cartographer Oronteus Finaeus (also known as Oronce Fine), showed a landmass startlingly similar to Antarctica. While the Oronteus Finaeus map puts the continent too close to the tip of South America and is oriented incorrectly, many of the other features of Antarctica appear in the right places, including an unnamed gulf hauntingly reminiscent of the Ross Sea. Equally important, the Oronteus Finaeus map shows a range of coastal mountains that now lies under the ice cap. This, too, is an ice-free Antarctica, one that boasts rivers and inlets where now only glaciers and ice shelves are found. To Hapgood it appeared obvious that Oronteus Finaeus, like Piri Reis, had access to ancient maps showing the southernmost continent in the long-gone days before it was capped by ice.

Hapgood also determined that these ancient mariners were sophisticated navigators. The Piri Reis map displays a web of lines, called rhumb lines, crisscrossing the Atlantic. Most scholars believe that rhumb lines, found on other maps from the same period as well, do not show latitude and longitude, but served as guides for laying a course by compass and wind direction. Hapgood begged to differ. Working with a mathematician from the prestigious Massachusetts Institute of Technology, Hapgood and his team developed a complex mathematical argument to show that the rhumb lines actually indicated longitude and latitude based on a projection with Cairo as the center point. This was a remarkable conclusion, since longitude couldn't be calculated accurately until the development of spherical trigonometry in the late seventeenth and eighteenth centuries, over 150 years after Piri Reis drew his map.

Hapgood also investigated a number of late medieval and early Renaissance maps known as portulans or portolanos. The accuracy of these charts, made by navigators with nothing but a compass, struck Hapgood as too close to perfect to be accidental.

From all this research Hapgood drew a number of significant conclusions. The Piri Reis, Oronteus Finaeus, and portolano maps showed geographical features that remained "undiscovered" until well after they were drawn, and they were surprisingly accurate. This would indicate that a civilization unknown to us and dating back to a time so ancient that Antarctica lacked an ice cap had mapped much of the world with a sophistication that proved the existence of advanced navigational aids and complex mathematics.

Hancock adds another candidate map to Hapgood's list, one drawn by the eighteenth-century French geographer Philippe Buache to show the globe from the South Pole up, as of 1737. Hancock claims this map is based on sources thousands of years older than the Oronteus Finaeus chart, although he doesn't say why he believes this. He also maintains that the Buache chart, like the Oronteus Finaeus map, provides an eerily accurate representation of how Antarctica must have looked when the continent completely lacked ice.

Drawing on Hapgood's work, the Flem-Aths conclude that Antarctica was more than simply known to the lost civilization posited by Hapgood. It was actually based there. Antarctica and its ancient residents are the lost place and people Plato called Atlantis and the Atlantians.

Like Hancock, the Flem-Aths add another key map, the previously mentioned work of Athanasius Kircher. Superficially it would seem that Kircher did little to support the Flem-Aths, since his map places Atlantis in the North Atlantic, between Europe and North America. Look deeper, the Flem-Aths argue. The Atlantis landmass has the same appearance as an ice-free Antarctica if you reverse the usual mapmaker's convention and make south, not north, up. The Flem-Aths argue that this indicates Kircher got his information from ancient Egyptian sources, since in Egypt south, the direction from which the Nile flows, was "up."

The Flem-Aths hold that approximately 11,500 years ago, during the time period Plato ascribes to the war between Atlantis and Athens, Antarctica wasn't where it is now. As a matter of fact, the whole crust of the Earth wasn't where it is now. In this prior incarnation of the Earth's surface, the Palmer Peninsula of Antarctica projected well into the temperate zone. This extension of the Antarctic landmass would have been warm enough to support a bustling civilization, while the rest of the continent was frozen and iced over. Then, within a relatively short period of time, the entire crust of the Earth slipped about 30 degrees, or approximately 2,000 miles—the distance between Minneapolis and Guatemala City—moving like the skin of an orange sliding over the sections beneath. This global crustal shift drew the Palmer Peninsula well into the Antarctic deep freeze, changing the climate in a cataclysmic direction. In the northern hemisphere, meanwhile, the movement of the polar ice caps into warmer southern regions initiated a rapid meltdown that raised sea level and slowly swamped the Atlantian cities. This took longer than Plato's single day and night, but its effect was no less catastrophic, sinking Atlantis forever beneath the waves.

Checking the Facts

The model advanced by Hapgood, the Flem-Aths, and Hancock comprises a series of hypotheses, one resting upon another like the stones of an arch. Supporting them all are three fundamental questions: Was a significant portion of Antarctica ice-free in circa 9600 B.C.? If the ice were removed from the Antarctic, would the continent assume the shape shown in the ancient maps? And are the maps as startlingly accurate as these authors say they are?

The Antarctic ice cap is the largest on Earth, covering some 98 percent of the continent to depths of three miles and representing 90 percent of the planet's total ice. The Antarctic ice cap is also Earth's oldest. The glaciation of Greenland dates back 7 million years. Antarctica was probably glaciated, at least in part, as long ago as 35 million years.

But even old ice caps may not be stable at their edges. If the climate changes toward more cold, an ice cap can expand its area by growing at the margins. Likewise, if the temperature warms to the melting point, it can retreat, shrinking back to expose previously covered land.

A current academic debate concerns just what has happened to the Antarctic ice cap in its 35-million-year existence. One group of scientists, led by Peter Webb, an Ohio State University geologist, has found evidence it considers indicative of a once-warmer climate, particularly bits of wood identified as southern beech, as well as beetle remains and tiny marine fossils left behind when the ice cap melted and ocean water flooded Antarctica's subglacial basins. Another group of scientists disagrees with this interpretation. Studying rock and ash samples from the same sedimentary layer that Webb investigated, George Denton of the University of Maine and David Marchant of Boston University found no evidence of any wetting or erosion that would have resulted from a glacial meltdown. Denton and Marchant argue that Webb's temperate fossils were deposited by the wind, and that Antarctica has been glaciated much as it is now for at least 23 million years.

Even if Webb's model is proved correct, however, it holds out little hope for the hypothesis of Hapgood, Flem-Ath, and Hancock. The temperate period Webb proposes occurred 3 million years ago, not 11,500. As of this writing, there is no credible evidence that any significant portion of Antarctica—certainly nothing the size of North Africa and the Middle East combined—was ice-free in circa 9600 B.C.

Another significant issue affecting the appearance of an ice-free Antarctica is the shape of the continent were the ice cap removed. Hapgood and the Flem-Aths assume that the ice would come off like the glass from a picture frame and in no way affect the image beneath. It doesn't work like that, however. The Antarctic ice cap is unimaginably massive, totaling an estimated 6 to 7 million cubic miles of ice and weighing in the vicinity of 30 million billion tons. This extraordinary weight, which is grounded on bedrock or stacked as ice rises on bedrock islands, pushes the continent down by hundreds of yards. Were the ice to melt away, the continent would bounce back—a geological process called isostatic rebound—raising its surface up to 3,100 feet in the interior and as much as 160 feet along the coast.

There's more. Melting of the Antarctic ice cap would spill a huge amount of water into the ocean and could raise sea level by an estimated 200 to 260 feet. The sea-level rise, combined with isostatic rebound, would produce an Antarctica markedly dissimilar in continental shape than today's landmass with the ice removed like that imagined pane of portrait glass.

It is this very issue of continental shape that opens the accuracy of the Oronteus Finaeus map to question. Hapgood himself admitted that a major problem in interpreting the map is the absence of the Palmer Peninsula, which trends out from the main Antarctic landmass toward South America for nearly 500 miles and would remain above the ocean surface even under the conditions of sea-level rise and isostatic rebound. This feature is so prominent that it would certainly appear on the most rudimentary chart, yet it is missing from the Oronteus Finaeus map. Likewise, the map shows the Wilkes Land portion of Antarctica as solid land, when in fact it comprises two large subsea basins with only an archipelago of islands rising above the surface of the frozen ocean. Were it ice-free, it would be an island-dotted embayment, not a solid mass. Similarly, the Amery Basin, which under ice-free conditions would be a 430-to-490-mile-long bay running perpendicular to the Antarctica coast, is missing. This kind of problem arises again and again. In other words, much of what the Oronteus Finaeus map shows as solid land would, in an unglaciated Antarctica, be underwater.

Oddly enough, the Oronteus Finaeus map may be a representation that is partly accurate and partly hearsay. This raises a tantalizing question about explorations of the southernmost continent that are lost to history. But if Antarctica is the landmass this map shows, it is the glaciated continent of modern time, not an ice-free version of the same landmass.

Similar problems plague the Buache map. It is no more successful than the Oronteus Finaeus at showing what Antarctica would look like in the absence of its ice cap. Simply, it doesn't show what Graham Hancock claims it does.

The Piri Reis map poses a more difficult set of problems, in part because of its age and style. How accurate is it? And does the map provide any internal evidence that it draws from sources dating to distant antiquity?

Actually, a number of significant details on the Piri Reis map have more of a medieval or Renaissance flavor than an ancient one. For example, the top of the map shows a large fish with two men sitting on its back and an inscription telling a tale from the life of Saint Brendan, the Irish mystic and seafarer who, legend reports, once said Mass on the back of a whale he mistook for an island. The original story, the *Navigatio Sancti Brendani Abbatis,* arose in Ireland before the eleventh century and is clearly medieval. So are the many islands scattered about Piri Reis's Atlantic Ocean. These hypothetical and imaginary bits of land are common features in late medieval maps of the vintage that Christopher Columbus used and Piri Reis likely borrowed from.

The Piri Reis map's version of the Caribbean shows further evidence of medieval origin. Only the eastern tip of Cuba is shown, a major omission, and Hispaniola, the island that contains today's Haiti and Dominican Republic, is oriented north-south rather than east-west. The likely explanation for this is that Columbus was certain that Hispaniola was really Japan, whose existence had been known since the time of Marco Polo and which was one of the desired destinations of the new route to Asia he was seeking. Columbus's own maps, like the one Kemal Reis had taken from the captured sailor, drew Hispaniola as if it were Japan according to the conventional representation of the time. Again, the origin of this part of the map is contemporary with Piri Reis, not an unknown ancient source.

Portions of the map are strikingly accurate, such as the Iberian Peninsula (Spain and Portugal) and the coast of west Africa, and the place names on these parts of the map are given in Turkish. The Ottomans sailed into these waters, and Piri Reis knew them firsthand or from accounts of his fellow naval officers. The coast of Brazil is also very accurate, with the relationship between Africa and South America much closer to reality than on European maps of the same time period. Place-names along the South American coast, however, are Turkish transliterations of Italian and Spanish, languages that arose from Latin only in the early medieval period. Most likely, Piri Reis drew this portion of his chart from accounts of the voyages of Amerigo Vespucci and other explorers, not from ancient sources written in classical Greek, Latin, or other "dead" languages.

Piri Reis wasn't the first cartographer to place a mountain chain in the place where the Andes later proved to be. One such map was drawn by Nicolo de Caneiro between 1502 and 1504. It survives today in the Bibliothèque Nationale in Paris, strongly resembles the Piri Reis map, and may have been one of the sources the Turkish captain used. He may also have referred to a similar map by Johannes de Stobnicza, who included his own charts in an edition of the Greek astronomer Ptolemy published in 1512. Since Ptolemy lived in the second century A.D., Reis may have mistaken Stobnicza's map for an "ancient" source.

And as for the creature Hapgood is certain can only be an Andean llama, it boasts two horns. Llamas don't have horns.

Ptolemy himself had referred to "a southern land" in his writing, and many of the explorers heading into the southern hemisphere kept an eye out for it. When Magellan sailed through the strait that now bears his name, he spied Tierra del Fuego to the south and thought he was looking at Ptolemy's southern landmass. Blown far off course, Amerigo Vespucci spotted land that may have been the Falkland Islands or perhaps even Antarctica itself. Other ships in similar situations reported similar sightings. As a result of all this, there was some contemporary, not ancient, knowledge of a body of land to the south of South America at the time when Piri Reis was drawing his map.

It's also possible that Piri Reis wasn't trying to draw Antarctica at all. One of the problems with the map, which Hapgood freely admits, is that about 900 miles of South American coastline are missing. Below the Rio Plata the coast turns to the east, and it is this east-bearing section that Hapgood posited as the coast of Antarctica. Yet, if the eastward coast is turned south instead, it represents a relatively accurate depiction of the eastern South American seaboard from the mouth of the Rio Plata to Tierra del

Fuego. Three islands shown on the map correspond well with the Falklands, and the tip of the continent appears just about where it should. And making this coast Patagonia rather than Antarctica explains Piri Reis's inscription about heat and big snakes.

In the end, the supposed mystery of the Piri Reis map could all boil down to practicality and scarce resources. Coming to the bottom of his valuable gazelle-hide parchment and running out of room to draw, Reis could have turned his coastline east rather than south. Mariners of the day hugged the coast whenever possible, and maps showing coastal shape were every bit as useful as ones with an accurate north-south orientation.

I continue to find Hapgood's ideas fascinating and tantalizing, but the evidence is less than cogent. The Piri Reis map contains no information of indisputably ancient origin, and the supposed coast of Antarctica could well be the lower reach of South America instead.

Yonaguni: The "Lost City" of Japan

In *Fingerprints of the Gods,* Graham Hancock argues in effect that Atlantis may be only the tip of an iceberg. In his view, there is ample evidence that a major civilization existed long before those first cities arose in the Middle East at the end of the Neolithic. The Inca empire and Machu Picchu, the mysterious lines on the Peruvian plain of Nazca, the Bolivian ruin of Tiahuanaco, the great Mayan temple sites like Uxmal and Chichén Itzá, Teotihuacan in Mexico, Giza's pyramids and Great Sphinx: these and many more sites tell Hancock that the Earth was once home to a major, highly sophisticated, and technologically advanced civilization that spread far across the globe before being suddenly destroyed by cataclysm.

Like Hapgood's theory, Hancock's is an intriguing idea, one that bears keeping in mind as we look at the mysteries of the past. When, not long after my work on the Sphinx, I received an opportunity to explore one of the most recently discovered sites pointing to the possible existence of this vanished and unknown civilization, I jumped at the chance to have a scientific look at it myself.

Yonaguni is the last island of the Ryukyu chain, which curves from Japan south and west toward China in the East China Sea. Part of Japan, as is its larger and better known neighbor, Okinawa, Yonaguni is the home of Kihachiro Aratake, a scuba-diving instructor and guide, and the head of the Yonaguni-cho Tourism Association. In 1987 Aratake was exploring the southeast coast of the island in a search for interesting new dive sites to offer his clients when he happened across something under the ocean's surface that took his breath away. Below him the submarine cliff appeared to be cut in a series of immense geometric terraces, their surfaces broad and flat, separated by sheer vertical stone risers. It looked, for all the world, like a reviewing stand for giant sea gods. Aratake felt sure he had discovered a structure fashioned by human hands.

By 1990, word of the Yonaguni Monument had come to the attention of Masaaki Kimura, a marine seismologist at the University of the Ryukyus in Okinawa. Interested in lost civilizations, Kimura dived at Yonaguni repeatedly over the next seven years, often bringing his students along. Kimura's team assembled a large, meticulous portfolio of drawings, maps, and models, not only of Aratake's discovery on Yonaguni, but also of other sites on the same island as well as others of the Ryukyus. Kimura himself became convinced that the Yonaguni formation, or monument, had been fashioned by human hands.

Japanese television and newspapers picked up on Kimura's ideas and gave them considerable coverage, a fact that brought the professor a great deal of ridicule from his academic colleagues. Yet, despite the hoopla in Japan, news of Yonaguni didn't leak out to Europe and North America until 1995 and 1996. So much for the global village.

I got wind of Yonaguni through John Anthony West, who himself heard about it from Shun Daichi, a translator who has worked with both West and Graham Hancock. Hancock, whose books have been runaway best-sellers in Japan, was very much intrigued by the Yonaguni Monument. News of his interest made its way to Yasuo Watanabe, an admirer of Hancock's work and a wealthy businessman. In 1997 Watanabe invited Hancock and his wife, Santha Faiia, to Yonaguni to have a look. Hancock and Faiia made several dives to the monument, whose regularity convinced them it was indeed the work of humans. Hancock, however, is no geologist, and he suggested to Watanabe that he bring John Anthony West and me over for a closer examination.

At first glance, Watanabe's invitation caught me off guard. I'm not keen on water, even less on diving in the open ocean, yet the chance of inspecting what could be a ruin of an antiquity equivalent to the Sphinx's intrigued me no end. Potentially, a great deal was at stake in Yonaguni. If it was a human-made structure, then it was reasonable to assume that it had been built when the site wasn't underwater. Based on well-established studies of rises in the East China Sea during and after the last ice age, the Yonaguni Monument was likely above sea level as recently as 8,000 to 10,000 years ago, or in the 6000–8000 B.C. range. Proof that Yonaguni is the result of human effort would mean that it is at least 8,000 years old and clear evidence of an early, perhaps unknown civilization. This possibility proved so intriguing that I sucked in my breath, responded with gratitude to Watanabe's generous offer, signed up for scuba lessons, and set off for Japan in September 1997.

Watanabe was a most beneficent patron. He brought not only John Anthony West and me to Yonaguni, but also Graham Hancock and Santha Faiia. On Yonaguni we were joined by Shun Daichi, several of Watanabe's American business associates, Kihachiro Aratake, dive masters Yoshimi Matsumura and Hiroshi-Kubota, and a four-person English documentary crew filming TV specials for Hancock's forthcoming book. Watanabe even lent us his own chef and, more a fisherman than a diver, kept us well fed with a steady supply of sushi from the snapper and tuna he hooked.

Diving off Yonaguni proved a challenge. The current was strong, the surge powerful, but with Yoshimi Matsumura as my personal diving partner and "assistant," I was in good hands. Probably I should have been more nervous than I was; sometimes naïveté falsely bolsters self-confidence. I have to admit, however, I kept looking over my shoulder for hammerhead sharks. These large predators are known to school off the island—a behavior that occurs in few other places in the world. Even though I was assured hammerheads usually leave humans alone, I imagined they might be interested in snatching an occasional scuba diver—like me—as hot lunch. Still, despite the difficulty and a constant low level of shark-dread, I was stunned by my first sight of the monument.

Superficially the monument has the appearance of a platform or part of a step pyramid, something like the ancient Temple of the Sun near Trujillo in northern Peru. The top of the monument lies sixteen feet under the surface, the bottom at an approximate depth of eighty feet. Extending over 160 feet in an east-west direction and more than sixty-five feet north to south, the asymmetrical monument has uneven stone steps, ranging in height from a foot and a half to several feet, on its southern face. It looks like a great staircase up which only a giant could stride. The surfaces have a regular smooth surface, like dressed stone.

In all, I made six dives on the Yonaguni Monument. Despite my initial shock, the closer I looked at the monument, the less convinced I was that Kimura was right about the structure's human aspect. Much of the regularity of the surface was due not to a tooled smoothness of the rock but to a thick, even coating of algae, corals, sponges, and similar organisms. In a number of spots I scraped the coating away, both to determine what kind of stone lay beneath and to look for tool scars or quarry marks. I found none. Even more telling, I couldn't find any evidence that

Yonaguni consisted of separate pieces of stone. Stone blocks carved, set in place, and arranged in an order would clearly indicate a human-made structure. Rather, the monument is essentially a single piece of solid, "living" bedrock that is less precise than it appears at first. The horizontal terraces aren't really horizontal, and the steps are not cut at precisely 90 degrees. A long channel about two feet wide, which looks as if it had been excised by some giant stone-cutting tool, proves to have a ragged, unworked bottom and to disappear into the bedrock as two parallel natural fault lines.

Still, Yonaguni posed a problem. If the monument was the result of a natural process, this natural process was unlike any I had seen before. What could it be?

Taking rock samples from the monument and inspecting the local geology gave me a tentative answer. The monument is composed predominantly of very fine sandstones and mudstones of the type we geologists call the Lower Miocene Yaeyama Group. Rocks of this type contain numerous, well-defined, parallel bedding planes that allow easy separation of the layers, and they are crisscrossed by many joints and fractures running parallel to one another and vertical to the bedding planes. Yaeyama Group sandstones lie exposed along the southeast and northeast coasts of Yonaguni Island, and I went there to see how they weather under current conditions above water. The more I looked at the highly regular yet completely natural weathering of these sandstones, the more I became convinced that the steplike and terracelike features of the underwater monument resulted from natural processes working on the stone, not from the activity of humans long ago. With its both vertical and horizontal fractures and joints, the stone was like a huge, many-tiered wedding cake cut into pieces and ready to serve. Subjected to thousands of years of surf, tides, typhoons, and storms, the rock had broken off in great square and rectangular chunks that left the visual impression of steps, plazas, and platforms.

Even though he was clearly more disappointed than I was, John Anthony West agreed. As a geologist, I was excited to have uncovered a new and unique geomorphology. West was on the track of a smoking gun pointing at the lost global civilization, and he had come up empty. As scientists working from the evidence, though, we could make no other deduction.

In late July and early August of 1998, I had the opportunity to return to Yonaguni under the auspices of Team Atlantis, a multidisciplinary underwater research group organized by Michael Arbuthnot. Studying the Yonaguni Monument for a second time, I remained convinced that the structure is primarily a result of natural processes.

Possibly the choice between natural and human-made isn't simply either/or. Yonaguni Island contains a number of old tombs whose exact age is uncertain, but that are clearly very old. Curiously, the architecture of the tombs is much like that of the monument. It is possible that humans were imitating the monument in designing the tombs, and it is equally possible that the monument was itself somehow modified by human hands. That is, the ancient inhabitants of the island may have partially reshaped or enhanced a natural structure to give it the form they wished, either as a structure on its own or as the foundation of a timber, mud, or stone building that has since been destroyed. It is also possible that the monument served as a quarry from which blocks were cut, following the natural bedding, joint, and fracture planes of the rock, then removed to construct buildings that are now long gone. Since it is located along the coast, the Yonaguni Monument may even have served as some kind of natural boat dock for an early seafaring people. As Dr. Kimura showed me, ancient stone tools beautifully crafted from igneous rock have been found on Yonaguni. Significantly, Yonaguni has no naturally exposed igneous rocks, so the tools, or at least the raw materials from which they were made, must have been imported from neighboring islands where such rock is found. The tools could have been used to modify or reshape the natural stone structures now found underwater off the coast of Yonaguni. The concept of a human-enhanced natural structure fits well with East Asian esthetics, such as the feng shui of China and the Zen-inspired rock gardens of Japan. A complex interaction between natural and human-made forms that influenced human art and architecture 8,000 years ago is highly possible.

Other evidence points as well to some kind of human working on the local stone on Yonaguni. Scattered over the island are apparently very ancient (age unknown), obviously human-carved stone "vessels." The vessels have no obvious use I could determine, and when I asked island natives about them, all I got were polite shrugs and blank stares. Fashioned from local rock and apparently neither carved nor transported to the island any time in the past 500 years, the vessels remain something of a mystery, as do the Yonaguni Monument and other underwater structures reported in the Okinawa area. At a minimum, the tombs, the ancient igneous-rock tools, and the vessels tell us that humans have been living and working stone on Yonaguni for a long time.

But if there is a human element to the Yonaguni Monument, why did these people of long ago want to mark this spot? Today the Tropic of Cancer, the northern boundary of the tropics, lies at 23° 27' north latitude. The Tropic of Cancer moves slowly on a 41,000-year cycle, for reasons we will discuss in chapter 5. Ten thousand years ago, in circa 8000 B.C., the Tropic of Cancer was in a different location, at about 24° 15' north latitude, which is close to Yonaguni's position at 24° 27' north latitude. Locating the exact position of the tropic is easy—all it takes is an exactly vertical stick or pole; a plumb line with a weight works nicely. The ancients, I suspect, knew where the tropic was, and they knew that, like precession, its position moved slowly. Since Yonaguni is close to the most northerly position the tropic reaches in its lengthy cycle, the island may well have been the site of an astronomically aligned shrine that, like Nabta in southern Egypt,

Stonehenge in England, or Brugh na Boinne in Ireland, replicates a celestial point on Earth and anchors our world in the eternal.

A possible parallel comes from ancient Egypt. The religious site of Philae, dedicated to the temple of the great goddess Isis, is located at 24° 1' north latitude, the position of the Tropic of Cancer 7,000 years ago, in circa 5000 B.C. The current buildings on the site date to Roman and Hellenistic times, but it is likely that they were erected on the ruins of much earlier structures. The Egyptian name for Philae means "Island in the Time of Re," a phrase that John Anthony West says indicates remote antiquity. Might the Egyptians, like the ancient people of Yonaguni, have built a monument to flag the position of the tropic?

In its way, Yonaguni, though not the smoking gun that West wanted, deepens the mystery. It adds, too, to the picture of an ancient people more sophisticated, more adept, and—yes—more civilized than convention has led us to believe.

The Martian Possibility

Some of the people interested in a lost civilization of long ago have ventured further than the last continent to be discovered or the edges of Earth's oceans. They have taken their hunt into space.

One school of thought, popularized in a number of books with only the most tentative scientific credentials, holds that civilization arose on Earth not as an indigenous event but as the result of contact with ancient astronauts. These space travelers ventured forth from their home planet at some point in the remote past to bring us the wonders of their own civilized and technological way of life. The best known of these writers is Erich von Däniken, whose numerous books have sold in the millions of copies and have been transformed into any number of sensationalist television specials with the air of a *National Enquirer* feature story and even less supporting evidence. Däniken has, for example, his own explanation of Hapgood's maps of a supposedly ice-free Antarctica as the result of contact with the continent not by ship but by space capsule. Likewise, the famed Nazca lines of Peru, which make sense only when viewed from above, are supposed proof of repeated overflight by space travelers of old.

A less televised example is Zecharia Sitchin, who makes the case that all of humanity was created by a group of aliens who visited this planet between roughly 450,000 B.C. and 13,000 B.C., a considerable reach of time. These space-traveling visitors genetically engineered the human race by combining their own DNA with that of early hominids in order to create workers for the mining enterprises they were founding on Earth (apparently, colonialism has existed in planetary worlds other than our own). Thus, just as the book of Genesis says, we humans were fashioned in the gods' own image. Sitchin proposes that the ancient Babylonian festival called the procession of Marduk commemorates the flight path the ancient astronauts followed in their journey through the solar system to a landing site on Earth.

In truth, I find this kind of thinking complete pseudoscientific blather. For one thing, it is fundamentally arrogant. Writers like Däniken and Sitchin trade on the notion that the development of civilization must be linear and that ancient people were clearly much more stupid than we are. Thus any sophistication, whether technological or philosophical, evident in the old ways must have been the gift of an outsider, since ancient people were too backward to have come up with it on their own. The truth, I suspect, is quite the opposite: Ancient people were much more sophisticated than we let ourselves believe.

The other obvious problem with the Däniken-Sitchin approach is the obvious lack of evidence for a civilization on another planet, whether in this solar system or another. That was why, when I first heard that some people were claiming to have seen evidence of ruined artificial structures on Mars in photographs taken by the *Viking* probe, I dismissed them as Däniken-

Sitchin babble. Then, as a scientist should, I looked at the data for myself—and I wasn't so sure.

The so-called Face on Mars was discovered by astronomer Tobias Owen on a photo frame of an area some thirty-four miles by thirty-one miles, taken at around 40° north latitude on the Martian surface, an area known as Cydonia. This same frame, snapped by the *Viking* Mars probe in 1976 from an altitude of 1,000 miles with poor resolution, shows other structures whose arrangement and appearance don't fit what we usually think of as natural landforms.

Vincent DiPietro, also an astronomer, discovered another image of the Face on yet another photo frame snapped by *Viking*. This photograph, taken thirty-five Martian days later than the first one and with different lighting, provided a comparative perspective and made measurement of the shape possible. The Face, which in the photos does indeed look something like a human face wearing a headdress or helmet, turned out to be about 1.6 miles long from head to chin, 1.2 miles wide, and about 1,200 feet high. Adding to the unnatural look of the Face were apparent lines that some observers took to be teeth in the mouth, presumed crossed lines above the eyes, and supposed lateral stripes on the headdress that bore at least a passing resemblance to the one worn by ancient Egyptian pharaohs.

Another interesting feature of the Cydonia region is the D&M (for DiPietro and his associate Gregory Molenaar) Pyramid, located about ten miles from the Face and aligned almost perfectly north-south along Mars's axis, much as the Great Pyramid of Khufu (Cheops) is aligned. A number of other seemingly artificial features in the same region have also been identified and named. There is the Fort, which has two distinctive straight edges; the City, an arrangement of massive structures interspersed with smaller pyramids; the NK Pyramid, about twenty-five miles from the Face and on the same latitude as the D&M Pyramid; and the Bowl, which is approached by a long ramp reminiscent of the stairways on Mayan and Aztec pyramids. Given that all these features are found in one relatively confined area adds to the enigma. Can it be by chance that so many seemingly artificial shapes are found so close together? Doesn't this fact alone speak to the presence of a guiding, intelligent hand?

At first I dismissed the idea of a ruined civilization on Mars as science fiction, the kind of thing that would come from an H. G. Wells or a Jules Verne. Then Mark Carlotto, a scientist who has done considerable serious work on the Martian images, let me examine a copy of them firsthand. Seeing is believing, and I found the photos surprising. The D&M pyramid showed what geologists call slump features on one side—the shape left when a muddy hillside, for example, slips away—but the regularity of the shape was most unusual for a natural landform. Even though I was not convinced that the Cydonia shapes were anything other than natural, there wasn't a good explanation for how and why they looked the way they did. Reluctantly I admitted to myself that the possibility of artificial structures on Mars deserved serious scientific scrutiny and investigation.

My reluctance arose from extreme skepticism about evidence for extraterrestrial intelligence. UFOs, extraterrestrial intelligence, and, for that matter, lost ancient civilizations have tended to act as magnets for various true believers who concoct elaborate theories of official conspiracy and occult connections across the millennia. The Cydonia mystery has certainly fit the pattern. The Internet has hosted any number of Web sites that propose a secret geometric relationship between the Cydonia structures and the Giza plateau, accuse NASA of conspiring to obscure the "truth," and even claim that ancient Egyptian cult practices are being secretly and purposefully used in the United States space program. Such nonsense gets in the way of the rigorous, dispassionate scientific work needed to get to the truth of the matter.

If Cydonia was the result of work by intelligent beings and included a human face and other Earthlike structures, two basic explanations were available. The first was largely a rewrite of Sitchin's astronauts-of-old idea—namely, that a civilization had arisen on Mars and come to Earth, perhaps with unkind intentions, and altered the history of our planet, possibly by importing technology. This notion I found hard to swallow, for the very good reasons I have already outlined. There was, however, a second basic explanation. According to this hypothesis, life arose first on Mars. Being farther from the sun, Mars cooled faster after the formation of the solar system and may have offered conditions favorable to the evolution of living things sooner than did Earth. Then, through some unknown agency or mechanism, living cells were transported from Mars to Earth, essentially seeding our planet. Perhaps the transmission went the other way, with life arising first on Earth and being transported to Mars. Thus life on the two planets shares a common origin, yet over the eons has followed evolutionary paths that were separate but in some ways convergent.

This second idea has a distinguished pedigree. One of the giants of chemistry, Svante August Arrhenius, who won the Nobel Prize in 1903, speculated that life on Earth arose originally on some other planet or planets and later spread here, as it did to other localities in the universe. This idea, sometimes called panspermia or the cosmozoic theory, has been discussed in various scholarly circles, but scientific evidence for the model has been lacking.

Or at least it was until 1996, when a NASA scientific team led by David McKay published a paper in the prestigious journal *Science* arguing that a meteorite found in Antarctica and generally considered to have come from Mars contained compelling evidence of microbial life on the Red Planet. The meteorite, technically named ALH84001, is a coarse-grained igneous rock that weighs a little over four pounds, is 4.5 billion years old, and was fractured and penetrated by water between about 4 and 3.6 billion years ago. Approximately 16 million years in the past, the rock is thought to have been ejected off the Martian surface as the result of an impact. It eventually made its way to Earth, where it fell as a meteorite approximately 13,000 years ago. In their studies of the meteorite, McKay and his colleagues found tantalizing evidence of the presence of life in carbonate minerals deposited in cracks in the Martian rock about 3.6 billion years ago. Some of the evidence was chemical, some of it shapes that suggested microscopic fossils of very small bacterial organisms. Although no one finding was itself definitive, the data taken as a whole led the NASA team to "conclude that they are evidence for primitive life on Mars."

This single sentence in the published paper caused an immense scientific stir and fired up the media. Any number of researchers attacked the findings as impossible, generally saying that the fossils were too small to be the remnants of living forms and that the chemical evidence resulted from biological contamination on Earth, not from life on Mars. Yet none of the objections fully undercuts the hypothesis, and new findings tend to support it. Further work by the NASA team shows that ALH84001 contains an abundance of hexagonal magnetite crystals that have a form produced by bacteria in anaerobic, or oxygen-free, environments and that match crystals found in Earth microbes in terms of size, shape, chemistry, and lack of defects. Again, the evidence is not definitive. Yet, as of this writing, the hypothesis that simple organisms similar to bacteria existed on Mars 3.6 billion years in the past is still alive and kicking.

Obviously, it is a long way from a bacterium to the Pyramid of Khufu (Cheops), yet the very suggestion that life could exist on Mars added an extra edge to interest in the Face on Mars and the other forms. Apparently yielding to pressure to provide additional data on Cydonia, and rising to accusations that the agency had been suppressing information about Martian cities and altering its photographs of the planet to cover up the evidence, NASA dispatched the *Mars Global Surveyor* to fly over the controversial area and take a new series of photos. These images, snapped in 1998 at ten times the resolution of the original *Viking* photographs and under different conditions of light and shadow, stripped away much of the Cydonia mystery. Now the Face looked less like a face than a weathered landform. As physicist John Brandenburg put it, "At first blush, the reaction is 'Gee, it doesn't look dramatic like the other pictures.'"

The initial *Viking* photos were taken in the Martian afternoon, when the low angle of sunlight lengthens and deepens shadows. The light also entered from the side, further distorting the Face. Imagine looking at yourself in the mirror in a darkened bathroom with a flashlight pointing toward your cheek. This was the perspective of the *Viking* images, with one side in light and the other in shadow. And since *Viking* wasn't directly over the Face, the pictures were taken at an angle through a cloudy atmosphere that blurred many details. By contrast, the *Mars Global Surveyor* photos were taken during the Martian morning with straight-on light. These conditions turned the Face into something more like a hill composed of different strata of rock that have weathered unevenly.

The more I have looked at these new images, the more I am convinced that they are quite interesting but entirely natural features. Arguing that they are artificial is a stretch, one simply not supported by the evidence we have. Based on what we now know, the Face and the other Cydonia structures are natural features formed by geological processes under Martian conditions, which differ from those on Earth and produce shapes of an appearance other than what we are used to.

Mars may indeed have once harbored life, and it may still. The recent discovery of colonies of bacteria in lakes of solid ice near the South Pole shows that microbial organisms could possibly exist under the frozen conditions found on Mars or the moons of Jupiter. This avenue of investigation should be pursued. Yet there is no reason to believe that the Atlantians arrived from Mars, in Plato's time or in any earlier era.

Bringing Atlantis Back to Earth

If Atlantis isn't located on Thera, the Antarctic, Yonaguni, or Mars, then just where is it? The answer, I suspect, is both prosaic and fascinating, literary and historical—or, more properly, prehistorical. It deepens, too, our understanding of the sophistication of ancient peoples.

It begins in understanding what Plato, the sole ancient source for Atlantis, was up to. To begin with, Plato was a philosopher, not a historian. He was interested in understanding the root nature of reality, not in providing a detailed and objective account of events in the distant past. In fact, the classical Greeks didn't conceive of history as we do. The very notion of recording history came into being only in fifth-century B.C. Athens with Herodotus, who died about the time Plato was born, and Thucydides, who died while Plato was still a young man. Before them, history was as much myth as fact, and it was recorded in the same poetic form as mythology. Even Herodotus and Thucydides, who chose prose over poetry as their mode of expression, picked up rumors and passed them on without much analysis, seeing their role as one of instruction and entertainment as much as one of weighing the facts. A good story remained a good story, a tale that deserved to be told and listened to. The pleasure and insight it gave mattered as much as its veracity.

Plato drew from this literary context. His dialogues are written in the form of a play, itself a form of fiction that allows considerable liberty with fact. They are not autobiographical; Plato does not participate in the dialogues. Rather, he writes in a third-person narrative about events set decades earlier, working much as a modern historical novelist would. And even within the *Critias* and the *Timaeus,* Plato takes pains to show how tenuous is the line of descent by which the story came to him. Supposedly the tale of

Atlantis was heard from Egyptian priests by Solon, the great sixth-century Athenian lawgiver who, after his retirement from public life, went to Egypt to study. Solon, according to Plato, wanted to turn the story of the war between Atlantis and Athens into an epic, an artistic creation of the same sort as the *Iliad.* Never completing the poem, which has not survived (if indeed it existed), Solon told the story to Dropides, who later told it to his son Critias, who, as an old man of nearly ninety, repeated it to his grandson Critias, then a boy of only ten, the one whose name attaches to Plato's dialogue. By the time Plato, who is Critias's great grandson, received the story, it had crossed seven generations, often from the very old to the very young. There is good reason why Critias invokes the aid of the goddess of memory when he begins his tale. It would be no small wonder if details or even major events had not somehow slipped away or been altered in the story's migration from mouth to mouth across so much time.

Yet Plato has Socrates make the point that the story of Atlantis "has the very great advantage of being a fact and not a fiction." Surely philosophers as insightful as Socrates and Plato understood the tricks memory can play with the mind over time, and the changes that make their way into stories told from generation to generation. What possibly can Plato mean by "fact" in this context?

I suspect that Plato was looking for the deeper level of philosophical fact he was always seeking. In Plato's view of the universe, objective data such as numbers and dates meant little. They were simply, in his famous metaphor, the shadows on the wall of the cave. What mattered to Plato was the light streaming in from outside. All his philosophical work was an investigation into its source.

One of Plato's abiding concerns was the nature of the perfect state. He believed strongly that the ideal political entity should be organized in a manner that corresponded to the universal forms underlying all reality. This interest led Plato to write the *Republic* and the *Laws,* both of which are concerned solely with just and right politics, and to spend some years in Syracuse as a tutor to that Greek city's ruler and an adviser in the pursuit of the perfect state. The *Timaeus* and the *Critias* belong to this theme in Plato's work. In both of them, Atlantis is discussed only as an example serving Plato's general concern with ideal political order. This connection is so strong that the renowned classicist Francis Cornford argued that together the *Timaeus,* the *Critias,* and the *Laws* amount to an unfinished trilogy on politics.

Yet, however freely Plato interpreted his poetic license in telling the story of Atlantis, there is an undeniable historicity in the two dialogues. Details of the tale have their roots in history well known in Plato's time. The bull worship of the Atlantians, for example, is similar to what we know about the worship and rites of the Minoans. Less than fifteen years before Plato wrote his two dialogues, the Peloponnesian cities of Bura and Helike were leveled by an earthquake and the latter sank into the Gulf of Corinth, a catastrophic event that may have given the philosopher an idea or two about how to destroy Atlantis. The staggering size of the Atlantian army sounds strikingly similar to Thucydides' overblown accounting of the Persian horde Xerxes massed against the Greeks. And the steadfastness of the Athenians against the Atlantian threat is much like the Athenians' willingness to stand, sometimes alone, against the encroaching and aggressive Persian empire that wanted to turn their independent land into just one more conquered province.

Plato's story of Atlantis is something of a mix-and-match. Some details he has likely invented, others borrowed from various historical events. As a result, there may well be a core of historical truth underlying Plato's brief accounts, one that teaches us something we didn't know about our past and that gives further insight into the effects of natural catastrophe on history.

Mary Settegast, in *Plato Prehistorian,* makes an argument I find both plausible and persuasive. She shows that in its broad outlines, Plato's story tells about a terrible war that rocked the Mediterranean world and displaced whole populations, including the first Athenians. And serving as a backdrop to this terrible war was a dramatic change in climate.

In Plato's telling, the god-fathered Atlantian race of 9,000 years before, or circa 9600 B.C., that wrote laws, tamed and raced horses, and controlled the land inside the Pillars of Heracles as far as Egypt and Italy began to lust for power and decided to attack the peoples of Europe and Asia. This description, Settegast argues, fits the Magdalenian culture, the same one that gave rise to the magnificent paintings of Lascaux and other cave sites in western Europe. A strikingly wealthy Paleolithic culture with a developed sense of artistic expression, the Magdalenians also apparently tamed horses—making them, as far as we know, the first Europeans to do so—and may have had a basic system of writing, attributes that Plato ascribes to the Atlantians. In its heyday the Magdalenian culture reached across most of western Europe as far as Italy. By 9600 B.C., Plato's date for the war between Atlantis and Athens, the Magdalenian culture was showing signs of decay. In its western reaches in particular, the Magdalenians were making more weapons than art, a sign that ambition and desire for power were replacing the esthetic pursuits of former days. Again, this fits Plato's description of an Atlantian people in decline.

Apparently the Magdalenians were acting on these new desires. In circa 9600 B.C., peoples of uncertain origin who were adept at both art and technology began settling what is now Israel, Palestine, Lebanon, and western Syria. Certain artifacts from these ancient sites, which are called Natufian, show a connection with early cultures in the Balkans. Cemetery sites to the north, in the Ukraine, and to the south, along the Nile, demonstrate considerable violence; many of the skeletons carry flint arrow- and spearheads planted deep within the bones, evidence that the dead were victims of war. Palestine, however, was peaceful, a haven from the conflict to the north and the south. It could well be that the Natufians, with ties to southeastern Europe, were refugees from the Atlantian/Magdalenian armed advance, perhaps the proto-Greeks who Plato said organized themselves against the threat from the west.

About a millennium later, in circa 8500 B.C., northern Europe and the Levant are littered with arrowheads and other signs of all-out war. This is the time period of Jericho's fortified wall and battlement tower, built to meet some external threat. And cave paintings from this period depict scenes of battle: massed warrior hordes, victims pierced by multiple arrows, combatants dancing in victory over the slain. The war ran its course over approximately five centuries, with the victor and the vanquished losing their identities as such over time.

Yet the cult of the warrior remained long after the hostilities ended. In much of the Middle East, including Çatal Hüyük, weapons were crafted with extraordinary care despite the absence of organized fighting and used at ceremonial sites. Perhaps these daggers, arrowheads, and spear points commemorated the great war of the ninth millennium B.C., even as Homer's *Iliad* memorialized the Trojan War centuries after the event.

Memory of the war against Atlantis could also be the source of the mythological story of Zeus's battle against the monster Typhon, a tale generally considered to have originated in the Middle East. In the tale Typhon attacks Zeus, who finally defeats him with sickle and thunderbolt. Many of the events in the myth as told by Apollodorus occur at geographical locations that are the very places where battles of the war of the ninth millennium B.C. raged. And the particular weapons Zeus used—the sickle and the thunderbolt, actually a type of polished stone ax—appear in abundance in these same areas.

Key aspects of this series of archaeological events fit Plato's outline. The western Mediterranean was pitted against the eastern, even as the Atlantians outside the Pillars of Heracles massed against the Athenians and their European and Asian allies. In Plato's telling, Africa was not attacked by the Atlantians. And in fact, even as hostilities raged in Europe and the Near East, Africa west of the Nile Delta was occupied by a different culture and free of war. The Magdalenian association with horses also fits, since the Atlantians were said to be descendants of Poseidon, god of horses, and they bred and raced the animals.

There is even a catastrophe, though on a time scale far longer than Plato's single day and night of destruction. As we have already discussed in relation to the ancient cultures of the Nile, the melting of glaciers and ice caps at the end of the last ice age raised sea level dramatically, with the Mediterranean rising two hundred or more feet beginning in the ninth millennium B.C. The swelling waters swamped many coastal settlements, displaced whole villages and perhaps even cities, and created a tide of homeless refugees seeking new places to live. The changing climate of these years resulted, too, in vastly increased rainfall. Archaeological evidence from the Middle East indicates that in about 7500 B.C., extraordinarily heavy rains resulted in widespread, catastrophic flooding throughout the region—perhaps the deluges that Plato says washed away the Athenian acropolis and stripped soil and forests from the Greek mountains. Such events may also have contributed to the war and its devastating aftermath brought on by Atlantian ambition, as the have-nots set upon the haves, and those with a lust for power exploited the weakness of their uprooted neighbors.

The tale that comes from Settegast's reconstruction of the archaeological evidence lacks the sudden drama of Plato's picture of a civilization destroyed in a single day and night of misfortune and thereafter lost to history. It does, though, convey evidence showing that the peoples of ancient Europe and the Middle East, like the culture that carved the Great Sphinx of Giza in predynastic Egypt, were accomplished and sophisticated.

The war of the ninth millennium B.C., which pitted the western

Mediterranean against the eastern, was as extensive and terrible in its day as the Hundred Years' War following the Reformation, or either of the two world wars of this century. Engaging in hostilities of such an extent requires great organizational and social skill. Armies, in Neolithic times as in our own age, must be trained, armed, transported from place to place, and supported by a logistical network. These people of long ago were engaging in major military endeavors long before the rise of anything we think of as a military kingdom or an empire. That fact says a great deal about their abilities.

Further evidence of the highly organized nature of the Magdalenians in particular comes from the striking homogeneity of that culture's art. Although tribal peoples in our own and recent times create art that varies from one tribe to another and from place to place, Magdalenian art was remarkably consistent throughout the whole of southwestern Europe. Whenever the style of the art changed, it changed across the whole culture.

There is more going on here than one artist copying his or her neighbor. It is rather as if some single source of authority were prescribing how art was to be done and promulgating these rules to all Magdalenian artists—something like the pope in the Vatican communicating points of the faith to local dioceses across the world. "Wherever it [the central point of control] may have been located, this was, following Plato, the home base of the Atlantic governors, from whom the art of Europe as far as Tyrrhenia was receiving its uniform direction and style," Settegast writes. Artistic uniformity, like the massive size of the ninth-millennium war, adds further to the impression of a large, well-organized people spread over an extensive area, yet able to act in an organized, unified, and coordinated manner.

The aftermath of the war, the rise in sea level, and the cataclysmic floods created a mixing of cultures that resulted in the striking brilliance of Çatal Hüyük. In its day, Çatal Hüyük was like the imperial Rome of the Caesars, a cosmopolitan center that attracted all of the religious traditions, cultures, and peoples of the time. Human remains in the city's ruins show that the inhabitants represented European and Middle Eastern physical types, and its wall paintings depict black-skinned dancers who must be Africans. The ritual life of Çatal Hüyük likewise mixed elements that came from the Aegean, Persia, and Egypt. In addition, the city's cult represented a major step forward from the religious practices of earlier times. Formerly, people made offerings to the gods and goddesses as a way of guaranteeing good harvests and healthy livestock. Going beyond this simple quid pro quo, the people of Çatal Hüyük practiced a mystery religion based on a search for freedom for the human soul through transformation and metamorphosis, an idea that appeared in classical Greece in the Eleusian mysteries and flowered later in Mithraism, Orphism, and Christianity. As with military organization and artistic uniformity, the advanced religion of Çatal Hüyük provides further evidence of the sophistication of these ancient Neolithic societies.

This, then, was Atlantis—a great, terrible, and long-past war exacerbated by natural catastrophe and involving peoples of striking skill and sophistication. Its history serves to remind us, as Plato said, that "There have been, and will be again, many destructions of mankind arising out of many causes; the greatest of these have been brought about by the agencies of fire and water...."

Fire and Water

PAUSE A MOMENT TO CONSIDER AGAIN PLATO'S WORDS: "THERE have been...many destructions of mankind...; the greatest of these have been brought about by the agencies of fire and water...." The ancient philosopher had a point. Again and again in human history, flame and flood have visited widespread destruction on civilization.

As an example of the role of fire, we can journey 3,200 years into the past. And for a sense of the power of water to change history, we can turn to the mythological tradition and the stunningly dramatic geological testimony of eastern Washington's water-scarred landscape. Then we must ask ourselves how such events come to pass.

The Bronze Age's Fiery Finish

Troy, whose destruction at the hands of vengeful Greeks Homer described in intimately gory detail in the *Iliad* and the *Odyssey,* was in fact but one of scores of cities and palaces razed at the end of the Bronze Age, circa 1200 B.C. Indeed, two successive Troys, Troy VIh and Troy VIIa in archaeologist's shorthand, were destroyed at the site known today as Hissarlik. The survivors of the first destruction moved back into the ruined city, which had been burned to the ground, and rebuilt it, only to see it leveled by fierce flames yet again.

The same fate befell practically every significant community in the eastern Mediterranean outside Egypt and Mesopotamia between approximately 1200 B.C. and 1175 B.C. No important Bronze Age site in Anatolia, within modern Turkey, escaped destruction; all were burned. Cyprus's three principal cities were consumed by flames, possibly twice, like Troy. Ugarit, in what is now western Syria, was burned to the ground and never again occupied. The destruction descended heavily on Palestine and what became known as Israel in the Iron Age, particularly close to the coast. None of the palaces of Late Helladic Greece—such as Mycenae, the regal seat of Agamemnon, the mythic Greek king who led the fleet of a thousand ships to Troy and was slain by his wife and her lover on his return—survived long into the twelfth century B.C. It is hard to find any important sites in the affected regions, except in the interior of the southern Levant, that were spared destruction at least once. Even smaller communities that escaped burning were abandoned inexplicably, their residents fleeing to someplace new, someplace they must have hoped would be a safe haven against the storms of death and destruction raging all around them.

The cultural and political effects of the Bronze Age catastrophe changed the course of human history in the region. The palace-centered culture of Myceanean Greece, which Homer lauded in the *Iliad* and the *Odyssey,* was lost except to the memory of poets. The Hittite empire, which had brought peace and prosperity to Anatolia, collapsed. The Levant recovered more quickly, but it was centuries before the same level of urban civilization was again reached. And Egypt, though it escaped the flames, was attacked by wave upon wave of armed refugees and so weakened by these struggles that the New Kingdom soon came to an end. Taken in sum, the cataclysm marking the end of the Bronze Age was the greatest disaster in ancient history, an event more destructive, though less well known, than the collapse of the Roman Empire.

Still, for all the physical evidence of this rapid and widespread catastrophe, its exact cause remains uncertain. There are scattered signs and stories of war, not the least of them being Homer's epic work. Yet there is no evidence of external invasion, as there is for the war between the Atlantians and the Athenian-led peoples of the eastern Mediterranean, or of extended hostilities between the principal peoples of the region, such as the Greeks and the Hittites. And if the devastation resulted from a war waged to steal wealth from others, why was the destruction so complete? It makes no sense for invaders to burn to the ground the very cities they wished to possess and occupy as their own.

This is an anomaly: a widespread, sudden, fierce destruction without apparent cause or obvious explanation. And, like any good geologist, I begin my search for a cause in the workings of planet Earth.

Mountains of Fire

As volcanoes sometimes do, Vesuvius gave the residents of Pompeii a warning. On August 24, A.D. 79, the formerly quiet volcano literally blew its top. Ash and pumice gravel exploded in a tall mushroom-cloud column towering over 60,000 feet into the atmosphere, then fell to the ground. The pumice and ash posed little direct threat to human life, but the weight of the accumulating debris, which piled up over eight feet in places, collapsed roofs and rendered houses dangerous and uninhabitable. Forced out of their homes, most of Pompeii's 20,000 residents fled, leaving behind about 2,000 people who chose to stay. Next morning the volcano erupted again, sending what is known as a pyroclastic flow—a fast-moving, ground-hugging avalanche of extremely hot ash and gas—into Pompeii. The Pompeiians who had stayed behind died within seconds or minutes, their desiccated corpses buried in fine ash and awaiting discovery by modern archaeologists.

The eruption that buried and preserved Pompeii was not particularly big, devastating, or unusual, by either modern or ancient standards. The most destructive volcanic eruption in this century occurred on May 8, 1902, on the West Indian island of Martinique. A pyroclastic flow surging down the slopes of Mont Pelée burned the town of St. Pierre to cinders and left 29,000 dead. And the nineteenth century saw two huge volcanic eruptions in what is now Indonesia. Tambora, which erupted in 1815, was the larger of the two, yet since it happened in a remote part of the world, its effects were little documented. Krakatoa is the better known, largely because its explosion, in August 1883, came after the invention of the telegraph, which allowed news of the blast to be sent rapidly around the world. Almost the entire island of Krakatoa detonated in one massive eruption that is estimated to have been fifty times greater than the atomic blast over Hiroshima. The explosion was so loud that it awakened sleepers in their beds in Australia, and it was heard as far off as Madagascar, about 2,200 miles away. Relatively few people were killed by the explosion itself, because the area was thinly populated. But the blast caused tidal waves, or tsunamis, that reached up to 130 feet in height. Almost 200 villages along the coasts of Java and Sumatra were swept away, and some 36,000 people were killed.

Volcanoes can also kill and devastate if part of the mountain falls away during an eruption, forming a thundering river of mud and debris the consistency of concrete, moving at fifty miles an hour. Called a lahar, such a flow from the Nevado del Ruiz volcano in Colombia buried the community of Armero at the foot of the mountain under thirty feet of mud and killed 23,000 people in 1985. A much larger lahar swept from Mount Rainier, outside Seattle, to Puget Sound 5,600 years ago. Today about 100,000 people live on what was the path of that mudflow, in the shadow of the volcano considered the most dangerous in the United States.

All by themselves, however, volcanoes are no more destructive than earthquakes or tsunamis, which can level cities and exact an enormous toll on life and limb, but whose effects are primarily local. Volcanoes are distinct in that they can have powerful deleterious effects over regions of the Earth far, far away from the eruption.

Krakatoa is a good example. When the volcano exploded, it sent a gargantuan column of ash into the atmosphere that turned the area around the volcano night-black for over twenty-four hours and rose to an estimated altitude of fifty miles. From there it spread in the windstreams of the high atmosphere, streaking sunsets with vivid colors and turning skies murky. Within a year the ash made its way from the southern hemisphere into the northern. Because the ash layer reflected sunlight back into space, the weather in the affected regions cooled—a smaller-scale version of the nuclear winter scenario. Following the Krakatoa eruption, temperatures were abnormally cool for five years. After Tambora erupted, the summer of 1816 was so unusually cold that crops in New England failed because of killer frosts striking repeatedly in the middle of summer. The next two years were only a little warmer.

The summer of 1601 was equally nonexistent. In February and March 1600, Volcán Huaynaputina in the Peruvian Andes erupted. According to research by Shanaka da Silva of Indiana State University and Gregory A. Zielinski of the University of New Hampshire, the eruption voided at least six cubic miles of molten rock and blanketed southern Peru and portions of Bolivia and Chile—about 180,000 square miles—with volcanic ash. About one thousand people died, and it took over 150 years for the economy of the area to recover fully.

Complementary research by Keith R. Briffa and his colleagues at the University of East Anglia, in Norwich, England, studied the cool-summer effect by examining wood density in tree rings. During warm summers, growing trees lay down denser wood, making ring density a good marker for summertime temperatures—denser annual rings in warmer years, less dense in cooler ones. Studying ring density data from across the northern hemisphere, Briffa found that the summer of 1601 stood out like the proverbial sore thumb, surpassing even the Tambora summers of 1815–17 for unseasonable coolness. The Huaynaputina eruption was probably at least partly responsible, although additional undocumented eruptions in North America may have also played a role.

Geological history reveals other, even larger eruptions that likely had profound climatic effects. The Yellowstone caldera in northwest Wyoming marks an enormous volcanic event of approximately 600,000 years ago. The quantity of pumice and ash lofted into the atmosphere was six times that of Tambora, fifty times Krakatoa's. Currently we have no accurate measure of the chilling this massive eruption caused, but it must have been profound.

Clearly volcanoes can cause local catastrophes with widespread regional effects on climate. Still, in terms of finding the culprit behind the fiery destruction at the end of the Bronze Age, they are an unlikely, even impossible, candidate. There simply is no evidence for significant volcanic activity in the Mediterranean region at the right time. The Thera eruption at 1410 BC ± 100, discussed in chapter 4 in regard to Atlantis, happened too early, and it was too small to account for all of the devastation occurring around 1200 B.C. Whatever brought the Bronze Age to its end, it wasn't a volcano.

The Moving Earth

But could it have been an earthquake? The eminent excavator C. F. A. Schaeffer first suggested that a devastating series of earthquakes was the Bronze Age culprit, and his idea has been championed by several archaeologists since then. Certainly, the idea bears looking at. The eastern Mediterranean is seismically active, and major earthquakes are known to have affected the area in both recent and ancient times. During the great battle of Joshua against Jericho, for example, which occurred about two hundred years after the end of the Bronze Age, it may well have been an earthquake that brought the walls tumbling down.

The problem with the earthquake explanation is that it doesn't fit the data well. As Robert Drews points out in *The End of the Bronze Age*, earthquakes often damaged ancient cities, but in the absence of other factors, like a simultaneous armed attack, they very rarely completely destroyed them. One of the few exceptions occurred in 373 B.C. when a devastating earthquake hit the Gulf of Corinth in Greece. The city of Bura was leveled, and the city of Helike sank under the waters of the widened gulf—an event that, as we saw in chapter 4, may have given Plato ideas about the end of Atlantis. Generally, however, when earthquakes struck ancient cities, the survivors would rebuild their homes rather than abandon them, as happened at the end of the Bronze Age. In some ancient cities, it seems, earthquakes became a regular part of life. Constantinople (formerly known as Byzantium, and now the modern Istanbul), for instance, apparently experienced a major earthquake on average once or twice a century in ancient and Byzantine times, and tremors much more often. Still, the city survived and flourished.

Furthermore, earthquakes in ancient times were not associated with devastating fires, like the blaze that turned San Francisco into a smoking ruin in 1906. The raging fires that often accompany modern earthquakes are due to broken gas pipes, snapped electrical cables, and so forth. Of course, ancient cities lacked gas and electricity. An overturned lamp might cause a small fire that could destroy a single building, but even that appears to have been an unusual occurrence. Most ancient cities were built principally of masonry, with the result that small, localized fires did not easily spread over great distances. A study of several hundred earthquakes in the eastern Mediterranean region between 600 B.C. and A.D. 600 showed that not a single one was responsible for a citywide fire. Yet it is this very sort of major conflagration, repeated again and again, that typifies the end of the Bronze Age.

Clearly earthquakes, like volcanoes, did not finish off Troy or any of the other great Bronze Age cities.

The Ubiquitous Tradition of the Flood

The *Iliad* and the *Odyssey* tell not only of Troy's last days and the war's terrible aftermath, but also of a fateful contest between strong men. On the Greek side there stand Odysseus, a warrior skilled in all manner of contending, a strategist sneaky enough to develop the idea of breaching Troy's walls in a wooden horse; Agamemnon the general and his brother Menelaus, who wishes to avenge the theft of his wife, Helen; Ajax, a warrior with more brawn than brain; and Achilles, the most brilliant fighter of all, the one whose wounded pride nearly leads to the destruction of the Greeks. The Trojans were led by Priam, their king, and Priam's son Hector, famed as a tamer of horses and a warrior to match mighty Achilles.

The two Homeric epics also tell the story of a fierce conflict among the gods. Achilles himself was part divine and part human, the son of the mortal Peleus and the immortal sea nymph Thetis. The gods and goddesses of Olympus chose sides: Zeus, Apollo, and Artemis supporting the Trojans; Poseidon, Hera, Aphrodite, Ares, and Athena backing the Greeks. Indeed, without Athena, Odysseus would never have survived his ten-year journey home to Ithaca, or found the strength to reclaim his palace from the suitors eager to marry his wife, Penelope, and take control of his wealth.

The mythic element in the Homeric poems makes an important point about the Trojan War and about the wider destruction that formed a backdrop for this particular drama. The catastrophe of the Bronze Age was not simply the work of human warriors caught up in a lust for adventure, riches, and the spoils of war. It was also the playground of gods and goddesses, those forces in the world that stand far above the merely human and that often govern fate according to their will and whim.

The same interplay arises in another, and much older, mythic event of widespread destruction. Practically everyone knows the biblical story of the great flood told in the Old Testament book of Genesis. Not long after God created humankind, he became angry at the evil doings of men and women and resolved to cleanse the Earth of all life in a deluge. God chose, however, to spare Noah from the general destruction because of his piety and goodness. God instructed Noah to build a huge floating box, or ark (from a Latin word meaning "chest"), and to fill it with his family and the male and female of every species of animal. No sooner had Noah finished his carpentry than God loosed torrents of water upon the world, drowning everything that breathed and floating Noah's ark on the great wet blanket that covered the face of the world. Months passed before the waters drained away enough for Noah and his ark to strike dry land, supposedly on Mount Ararat in modern Armenia, and offer sacrifice to God, who promised never again to destroy the world by flood.

The Greeks told a similar story, detailed by Apollodorus and borrowed for retelling by the Roman poet Ovid. Like the tale of Noah, the Greek myth concerns the fate of humankind in the early days soon after the original creation. Zeus was angered by the impious sons of Lycaon, who sacrificed boys to the gods and who tried to deceive Zeus into eating their flesh. Zeus decided to remove all humans from Earth as well as Lycaon's blasphemous brood, and to this end he prepared a great, world-cleansing flood. Warned by his father, the Titan Prometheus, about the coming of the waters, the king Deucalion constructed a huge floating box and went aboard with his queen, Pyrrha, just as Zeus unleashed the flood. After the waters crested and subsided, Deucalion landed on a mountain—some say it was Parnassus; others claim it was Aetna in Sicily, Athos in northern Greece, or Othrys in Thessaly—and sacrificed to Zeus.

The stories of Noah and Deucalion share a great many similarities in both general outline and specific detail. The gods share similar motivation, and the mortals escape through a divine warning that tells them to build a big floating box. To determine whether the flood had really ended, Noah released a series of birds from the ark. When the dove returned with an olive branch in its mouth, he knew the worst was over. Deucalion was likewise reassured about the end of the cataclysm by a returning dove. Given these and the many other commonalities between the two myths, it is more than likely that the Noah story or a similar tale made its way from the Middle East to Greece, where it was revised to fit local gods and geography.

Yet the myth of Noah's flood hardly originated with the Hebrews. The scriptures for the early books of the Old Testament were recorded in their current form in the fifth century B.C., when the temple in Jerusalem was rebuilt after the return of the Hebrew people from captivity in Babylon. A much older Mesopotamian source, the *Epic of Gilgamesh,* recounts a flood story unmistakably similar to the Noah and Deucalion tales.

Gilgamesh, the part-human, part-god king of the city of Uruk, suffered inconsolable grief following the death of his beloved friend, Enkidu. Resolving to unravel the secret of life and death, Gilgamesh made his way through many dangers and obstacles to find Utnapishtim, a mortal man who had earned the gift of immortal life. In a story within the story, Utnapishtim told Gilgamesh of the world-destroying flood he had survived many centuries earlier. Again the cause was divine anger, in this case from the god Enlil, who was tired of having his heavenly sleep disturbed by the ceaseless clamorings of humans below. Warned of the coming catastrophe by another god, Ea, Utnapishtim built yet another floating box and filled it with living creatures in breeding pairs. Again the flood waters came, again they receded, again the hero learned of the return of dry land through the behavior of a bird, and again he sacrificed to the gods. Contrite over his destructive anger, Enlil bestowed eternal life upon Utnapishtim and his wife.

The final version of the Gilgamesh story dates to the seventh century B.C., yet its sources are considerably older. Most scholars believe the poem was originally written down by no later than the early centuries of the second millennium B.C. It certainly existed in oral form before that, and perhaps even in written form, which pushes its date of composition back into the third millennium, between 3000 and 2000 B.C. And if the epic commemorates a real event, the catastrophe that inspired the poem must have happened sometime before 2000 B.C.

Remarkably, this particular flood tale, which arose in Mesopotamia and then traveled to Israel and Greece, is only one of an extraordinarily large number of mythic flood stories. Among the Greeks, for example, the story of the deluge from which Deucalion was spared was not unique. Greek mythology recounts other floods of equal devastation. In the *Critias,* Plato tells of the deluge that swept away the Athenian acropolis: "For the fact is that a single night of excessive rain washed away the earth and laid bare the rock; at the same time there were earthquakes, and then occurred the extraordinary inundation, which was the third before the great destruction of Deucalion." If Plato is to be believed, apocalyptic floods were a regular horror in preclassical Greece.

As we have seen in our discussion of Atlantis, archaeological evidence from the Near East points to catastrophic flooding in the region in circa 7500 B.C. Bill Ryan and Walt Pitman of New York's Lamont-Doherty Earth Observatory have argued that another great flood occurred in the eastern Mediterranean in about 5000 B.C. At that time the Black Sea was smaller than it is today, and composed of fresh water, making it more lake than sea. A huge plug of silt separated it from the Mediterranean. Then, as the Mediterranean's sea level rose in the aftermath of the last ice age, salt water forced its way through, first as a trickle, then in a cascade with the force of two hundred Niagaras and a roar that could be heard for sixty miles in every direction. Within a few months, over 35,000 square miles of land were flooded, destroying farming villages on the shores of the former lake and possibly giving rise to stories that made their way to Mesopotamia and the poets who composed Gilgamesh.

If these ancient floods were the only such cataclysms of prehistorical times, we would expect flood stories to originate in the Near East and the eastern Mediterranean alone. Yet flood stories from traditions other than the Gilgamesh-Noah-Deucalion lineage appear literally all over the world. Accounts of floods so massively destructive that they threatened all living things are found in indigenous cultures throughout Europe, the Middle East, Persia and the Indian subcontinent, Tibet and Mongolia, China and East Asia, the Philippine and Indonesian archipelagoes, Australia and the Pacific islands, and North, Central, and South America. Flood stories number in the hundreds and span an extraordinary range not only of the globe but also of cultures, physical environments, climates, religious traditions, and language groups.

Such a welter of myths provides a body of fascinating if intangible evidence. Were there only a few such stories, we could dismiss them as exaggeration or imagination. Yet this global diversity of traditions, all of them telling of floods that threatened the very existence of our species, points to some terrible, long-past event or events lost to direct memory, yet preserved in the ancient form of storytelling.

There are two basic possibilities. Either an overwhelming flood devastated the Earth long, long ago, before humankind fanned out over the planet and carried the story of the destruction with it. Or sudden, terrible, seemingly world-destroying floods were a regular feature of human life, at least in prehistoric times.

The Great Floods of North America

Further evidence of the prehistoric reality of huge floods comes from our own continent. As the last ice age was ending and sea level was rising, portions of North America were swept by floods that, like the rush of the Mediterranean into the Black Sea, would have certainly looked like the end of the world to anyone who witnessed them.

Based on evidence taken from sediments in the Gulf of Mexico, C. Emiliani and his scientific colleagues argued that about 11,600 years ago a great rush of fresh water flooded down the Mississippi Valley from the Great Lakes region into the gulf. The water came from extremely rapid melting of the retreating glaciers in what is now the American upper Midwest and southern Canada. According to Emiliani, the flood was so great that sea level around the world rose by a foot or more a year—an astonishing one hundred feet in a mere century at that rate—which "could be an explanation for the deluge stories common to many Eurasian, Australasian, and American traditions." And, of course, the date of Emiliani's flood corresponds almost too perfectly with Plato's date of 9600 B.C. for the destruction of Atlantis.

Emiliani's work has drawn considerable fire from other geologists, notably Herbert Wright. Wright argues that the data don't support the suddenness Emiliani proposes. A great deal of fresh water did enter the Gulf of Mexico, but over a much longer period of time that stretched between 18,000 and 11,500 years ago. And, Wright says, geological evidence on the size of the North American glaciers, and especially the volume of water in the ice that melted during the millennia in question, indicates that the total amount of water added to the oceans from this particular meltdown may have amounted to only two feet over the whole 6,500 years, not a foot a year.

In evaluating these competing ideas, I am struck by Emiliani's overprecision and Wright's clear bias toward a uniformitarian explanation. Emiliani's scenario is questionable at best, but he may be correct in some general way, if not in the specifics. In fact, when we investigate ice ages later in this chapter, we will discuss a sudden warming trend that occurred at just about the time Emiliani hypothesized.

The great North American flood that is not subject to question by scientists and scholars occurred in what is now the Pacific Northwest, approximately 15,000 years ago. At the time, a huge inland sea called Lake Missoula, which had backed up behind an immense glacial dam in northern Idaho, extended far into western Montana and contained more water than present-day Lake Erie and Lake Ontario combined. When the dam collapsed, 500 cubic miles of water rushed out, beginning as a wave of water and ice 2,000 feet high, surging westward toward the Pacific at an estimated sixty miles per hour, and flowing at a rate ten times greater than all the rivers of Earth combined. Stop and think about that for a moment. If you put together the planet's rivers—the Amazon, the Congo, the Mississippi, the Plate, the Ganges, the Mekong, the Nile, the Saint Lawrence, the Danube, the Don, the Volga, and all the rest—and consider how much water they discharge in an hour, you would still have only one-tenth the per-hour quantity of water that poured out of Lake Missoula during the two days of the catastrophe. In its fierce rush, the flood carved canyons, scooped out enormous plunge pools below waterfalls that would have made Niagara look like a child's plaything, forced rivers such as the Snake and the Willamette to flow backward, cut the spectacular cliffs of the Columbia Gorge, and stripped away 200 feet of soil in parts of eastern Washington, leaving only bare basalt. The flood may have happened again and again, possibly dozens of times, as the glacial dam re-formed, held the waters back, then collapsed once more.

As the evidence from Monte Verde shows, the Americas were inhabited by humans when Lake Missoula flooded. An event so calamitous would surely make its way into mythology and story. And if such a flood occurred on this continent, others may well have occurred elsewhere in the world.

The Coming—and the Going—of the Ice

The ancient floods we have described—the torrential rains of circa 7500 B.C. in the Middle East, the Black Sea flood, Lake Missoula's high-speed rush to the Pacific, and Emiliani's questionable scenario of a sudden freshwater surge into the Gulf of Mexico—all share a common element: the retreat of heavy glaciers at the end of the most recent ice age. To understand more about the capacity of water to wreak devastation, we need to look at the climatic processes leading to the advance and retreat of ice caps.

To begin with, weather and climate aren't the same thing. To understand the difference, consider climate what we expect and weather what we get. To characterize climate, scientists look at averages of weather over time in terms of measures like inches of snowfall and rain, number of cloudy versus sunny days, wind speed, maximum and minimum temperatures, and similar data. Usually a very wet year, for example, is balanced out by a very dry year, so that climate tends to balance out around some relatively constant point.

Yet climate can vary over time, sometimes abruptly and fiercely. Geological history shows that over the past several million years there have been a number of periods, each lasting tens of thousands of years, when the climate over the entire Earth was much colder than it is now and when extensive glaciers covered significant portions of the northern hemisphere. During the past 750,000 years alone, major glaciers have extended from the polar region southward into the United States and Canada at least eight times. The ice fields reached their greatest size about 20,000 years ago, penetrating as far south as Ohio and Indiana and reaching depths of up to two miles. Then, starting approximately 14,000 years ago, the ice fields began to melt and retreat, uncovering the land that had lain beneath them and altering the topography and biology of the continent. Glacial melt filled immense basins gouged out by the ice, creating the Great Lakes, Lake Winnipeg, Great Bear Lake, and Great Slave Lake, as well as the tens of thousands of smaller bodies of water in the continent's center. The wet forested zone between the Rockies and the Cascades dried up, forming today's Great Basin deserts. The extensive waterways in that region vanished or shrank, leaving the Great Salt Lake—which once was 1,200 feet deeper than it is now, and as big as Lake Michigan—a shadow of its former self. Cold-loving spruce trees retreated a thousand miles north with the glaciers, and a number of animal species that had once been common during the glacial period, such as mammoths, ground sloths, and saber-toothed tigers, disappeared.

Although the ice has yet to return, by no means has the climate been constant since the end of the glacial retreat in circa 5000 B.C. For example, between A.D. 800 and 1200, the northern hemisphere warmed so much that wine grapes were cultivated in Norway, and Vikings pioneered farming villages on Greenland. In the 1400s the climate turned so cold that the Greenland settlers could no longer grow crops and had to abandon the island to the Inuit. The climate remained cold until the 1800s, forming what is sometimes referred to as the Little Ice Age. A dramatically beautiful reminder of this four-hundred-year period of glaciation can be seen in California's Desolation Valley Wilderness Area, southwest of Lake Tahoe. The wilderness area is named for a sparkling gray-granite valley scoured out by a glacier that lasted well into the nineteenth century. Only in the past few decades have a few hardy whitebark pines moved onto the granite the glacier scraped clean, leaving a landscape as stark as the moon's and as stunning as a Taoist silk painting.

Apparently climate moves in both very long-term cycles and shorter-term changes. A number of theories have been proposed to account for both types of climatic behavior. For example, short-term climate changes, like the Little Ice Age, have been attributed to sunspots. Sunspots are essentially storms or flare-ups of the surface of the sun that occur with some predictable periodicity on cycles lasting eleven and twenty-two years. There also appears to be a longer-term cycle in which sunspot activity drops to a minimum approximately every 2,500 years. The Little Ice Age does appear to correspond to the last period of low sunspot activity. The problem with this model is that it provides no clear mechanism connecting sunspot activity with the Earth's climate. Simply, what do sunspots have to do with climate? Increased sunspot activity might send more solar energy to the Earth's surface, but it is uncertain if this is the case, or whether the increase would significantly affect the climate. We just don't know whether there is any connection other than coincidence.

Plate tectonics, a theory of continental movement we will explore later in this chapter, has also been put forward as a model to explain the ice ages. Shifts in relative position between continents and ocean basins can drastically affect climate. But plate tectonics works very slowly, over millions of years, and it can hardly account for an event like the melting of the most recent ice age over a mere seven-thousand-year period. Very likely the continents have to be arranged in a particular pattern for an ice age to occur, but all by itself the model doesn't identify the cause of encroaching glaciation or its subsequent retreat.

Volcanoes have also been put forward as an ice age cause. As Krakatoa, Tambora, and Huaynaputina show, volcanic eruptions can put enough dust into the upper atmosphere to substantially cool the planet for three to five years. Yet three to five years do not make an ice age, nor are there good data to prove a convincing correlation between extensive volcanic activity and growing glaciers. Volcanoes may well contribute to glaciation, but it appears unlikely that they can initiate an ice age all by themselves.

To better understand the genesis of ice ages, we have to take a step back and remind ourselves of something basic: Earth is not all alone in the universe. The best explanation for warming and cooling trends in climate comes from understanding the subtle mechanics of our planet's orbit around the sun.

In terms of energy flow, the Earth's surface is an open system. It constantly receives energy from the sun and constantly returns this energy to space. Approximately 30 percent of the solar energy that hits the upper atmosphere is reflected back immediately, and another 20 percent is absorbed by water vapor in the air, leaving only 50 percent to reach the surface of the planet. Most of this 50 percent goes into evaporating surface water and thus reenters the atmosphere, from which it eventually radiates back into space as heat. This cycling of energy, from the sun to the surface of Earth and back into space, is the machine that drives weather and makes climate.

Obviously, if less energy is coming into the system or if more is radiated back, then Earth cools. Glaciers can form. And, equally obviously, if more energy reaches the surface or if less of it returns to space, the planet warms. Then any glaciers in existence can retreat or even disappear.

As far back as the nineteenth century, various scientists wondered whether Earth's orbital path around the sun and changes in the pattern of energy striking the surface of the planet had anything to do with the beginnings and the ends of ice ages. The mathematician Milutin Milankovitch elaborated and popularized this idea in the 1920s, developing a theory that most scientists of the time dismissed as wrongheaded. Data gathered since the 1970s, however, have shown that Milankovitch pretty well understood the process.

Milankovitch maintained that ice ages come and go through the complex interaction of three separate orbital cycles, all of them happening simultaneously. The first of these is called *eccentricity*. That is, Earth's orbit around the sun follows the path not of a circle, but an ellipse—which means the orbit is slightly oval. What's more, the ellipse slowly gets longer, then gets shorter again, varying about 2 percent approximately every 100,000 years. When Earth is farther away from the sun in its larger orbit, it receives less energy and cools, while it receives more and warms when it is closer.

The second cycle Milankovitch identified involves the tilt of Earth's axis. The axis on which our planet rotates isn't perpendicular to the plane of the orbit. Rather it is tilted, currently by approximately 23.5 degrees. This tilt, known technically as *obliquity,* varies between 21.6 and 24.5 degrees over a 41,000-year cycle. In the low latitudes closer to the equator, tilt affects climate minimally. But at high latitudes, particularly near the poles, the influence of tilt is profound. When Earth reaches maximum obliquity of 24.5 degrees, the winter dark extends farther south, into lower latitudes, and the cold is deeper. And when the obliquity moves toward the minimum of 21.6 degrees, the winter dark is confined to higher latitudes and the season itself is warmer.

Obliquity interacts with the third of Milankovitch's cycles, the one we have seen before: precession. This slow wobble of the Earth's axis affects the seasonal balance of radiation because it changes the timing of the points in the annual orbital cycle at which Earth is farthest from and closest to the sun. The timing shift moves on a cycle of about 22,000 years. Between them, tilt and precession can alter the amount of solar radiation in high latitudes by some 15 percent.

The strongest validation of Milankovitch's ideas has come from deep cores drilled from the Greenland and Antarctic ice caps. A cap forms when snow accumulates over tens of thousands of years, compresses under its own weight, and turns into ice, which contains physical, chemical, and even biological evidence of what was happening on the planet at the time the snow fell. The ice cores can, for example, inform us about Earth's general temperature through the millennia, and even provide samples of ancient atmosphere.

Thorough study of ice cores, combined with analyses of deep-sea sediments, indicates that temperatures on Earth were warm about 103,000, 82,000, 60,000, 35,000, and 10,000 years ago. These data fairly closely fit Milankovitch's precessional cycle of approximately 22,000 years and the tilt cycle of 41,000 years.

Were one a uniformitarian, such findings would offer philosophical comfort. In geological terms, of course, 20,000 years is little more than the slow blink of an eye, but from the perspective

The Giza Plateau, outside modern-day Cairo, Egypt. The Great Pyramid of Khufu (Cheops) lies on the right, and the Pyramid of Khafre (Chephren) is behind it to the left.

Looking north toward the Great Sphinx, which lies in the center of this view. The Valley Temple sits in front of the Sphinx and obscures the Sphinx Temple, also to the front of the Sphinx. The Great Pyramid of Khufu (Cheops) is in the background, along with the suburbs of Cairo—Egypt's ancient and modern faces.

A side view of the east-facing Sphinx, seen from the north.

In front of the Sphinx lie the ruins of the Sphinx Temple; behind it, the Pyramid of Khafre (Chephren).

Above left: Time has been hard on the Sphinx. The rolling weathering pattern on the body came from heavy rainfall pouring over the monument in ancient times. Weather damage on the paw and the lower body has been repaired with limestone blocks.

Above right: John Anthony West, left, and the author in front of the Great Sphinx.

Deep, rolling, rounded, and precipitation-induced weathering with distinct vertical crevices is particularly prominent in the Sphinx enclosure. John Anthony West faces the camera, with Jorjana Kellaway partially hidden behind him. Seismology expert Thomas L. Dobecki works below.

This perspective provides an additional view of the heavy weathering inside the Sphinx enclosure. Obviously caused by water, this pattern was probably left by the heavy rains of the Nabtian Pluvial, which began as early as 10,000 B.C. and lasted as late as 2500 B.C.

Top: Wind leaves a very different pattern of erosion than does water, as can be seen in this view of Old Kingdom tombs at Giza. Wind-driven sand scours out the softer portions of the rock and leaves the harder material behind.
Center left: Another example of weathering, caused by wind-driven sand.
Bottom left: The original limestone blocks of the Valley Temple were exposed to the same heavy rains as the Sphinx and, like it, reveal precipitation-induced weathering.
Bottom right: The entrance to the Valley Temple shows how granite ashlars, probably fashioned during the time of Khafre (Chephren) around 2500 B.C., were carved to fit the older, underlying limestone blocks contemporary with the Sphinx—part of the evidence that the Sphinx dates to well before the period of the pharaoh conventionally considered its builder.

Above: The author with the stela of Thutmose IV.

Above left: The rump of the Sphinx shows evidence of repairs both modern and ancient.

Left: The stela of Thutmose IV (also known as Tuthmosis IV), a New Kingdom pharaoh of circa 1400 B.C., sits between the paws of the Great Sphinx. The inscription originally contained one hieroglyph, now flaked off the stone and lost, that has been taken by Egyptologists to be part of Khafre's name and accepted as evidence that the Old Kingdom pharaoh built the Sphinx.

The lack of heavy precipitation-weathering of these mudbrick mastabas, which date to 2800 B.C. on the Saqqara Plateau is further evidence that the Sphinx's weathering pattern argues for an older date of origin. The structure in the background is the Step Pyramid of Djoser (Zoser), which was built circa 2650–2600 B.C.

Waves wash ashore on Yonaguni and form a rectangular stone block by natural erosional processes. The pattern of vertical joints and fractures common to this group of stones appears clearly in the rocks on the right.

The south side of the Yonaguni Monument. At first glance, the structure suggests one of the great monuments of the ancient world, complete with platforms and giant stairs, and seems as if it must have been the work of human hands.

Right: Dale Kimsey, left, and the author examine the Yonaguni Monument first-hand.

One of the fascinating tombs of uncertain age found on Yonaguni. Carved out of solid bedrock with an unmistakable resemblance to the Yonaguni Monument, this tomb measures approximately eighteen feet from base to top.

Further views of the south side of the Yonaguni Monument.

of human life span and civilization, it amounts to a long, long while. However, the Greenland ice data also show a number of events that don't fit Milankovitch's slow cycle. Rather, these changes are both rapid and unpredictable.

A particularly striking revelation from the Greenland ice cores is a series of what are called *interstadial events.* During these events, each of which lasted from a few hundred to a few thousand years, Greenland warmed quickly, then cooled, first slowly, then very fast. Apparently the change affected much more than just Greenland. During the interstadial events, the methane concentration in the ancient atmosphere increased markedly. Methane comes primarily from bacteria inhabiting oxygen-scarce environments, such as bogs and swamps. Apparently, when the climate warmed at the beginning of the interstadial events, tropical wetlands expanded, probably because of increased rainfall and a global warming trend. The warming trend reached as far as the South Pole. Recent research has shown that two of the sudden warming episodes revealed in the Greenland ice cores are mirrored as well in the ice of the Ross Sea area of Antarctica.

Interstadial events are hardly uncommon. In the period between 100,000 and 20,000 years ago, about two dozen of them occurred. And there was nothing reassuringly slow about them. In mere decades, the temperature warmed 9 to 18 degrees Fahrenheit, snowfall doubled, and an up to tenfold increase in dust trapped in the ice indicates stronger and stronger winds.

A recent study of an interstadial event beginning approximately 11,700 years ago shows how startlingly sudden these changes can be. The entire transition took about 1,500 years and represented a substantial warming of approximately 27 degrees Fahrenheit—a massive increase. Remarkably, the ice-core data suggest that half of the temperature change, in the neighborhood of 14 degrees Fahrenheit, occurred in less than fifteen years centering around approximately 9645 B.C. That's a bigger temperature increase, and faster, than the scariest doomsday scenario about global warming in the twenty-first century. Yet it happened well before humans were burning fossil fuels by the millions of tons and adding greenhouse gases to the atmosphere at unprecedented speed.

Two points about this finding are worth emphasizing. The first is that against the uniformitarian backdrop provided by the Milankovitch cycles, something much faster is also happening. The second is the stunning lineup in time between the sudden warming of 9645 B.C., Emiliani's scenario of a massive freshwater flood pouring into the Gulf of Mexico, and the date Plato ascribed to the sinking of Atlantis. Whatever the accuracy of specific details, this curious coincidence points to the effect sudden climatic changes can have—and no doubt have had—on civilization.

The Planetary Onion

The Milankovitch model tells us something, but not enough. Obviously some additional model is needed to explain those seemingly unexplainable interstadial events, with their striking abruptness and potentially disastrous consequences. To evaluate the ideas and theories that have been proposed, we need first to consider how our planet is assembled and what makes it tick.

If you had a knife big enough to cut a slice through Earth, you would see that the planet is formed of concentric rings, something like the layers of an onion. The outermost layer is the gaseous envelope of the atmosphere, which, although it extends out into space in rarefied form for hundreds of miles, is by far thickest in the twelve miles closest to the surface of the planet.

The uppermost layer of the solid Earth is the crust. Remove all the water from the surface, and you would observe the crust's two basic terranes*: ocean basins averaging about three miles in depth below sea level and floored with basaltic rocks, and continents and continental islands, which on average rise a few hun-

***In geology, a terrane is an area or surface over which a particular kind of rock is prevalent.**

dred yards above sea level and are founded on granitic types of rocks. The crust under the oceans is thin, averaging only about three miles thick, while the continental crust generally varies in thickness from about twelve to thirty-seven miles. There are differences, too, in the chemical composition of the crust under continents and oceans. The seabed crust comprises about 50 percent silicon dioxide (which you probably know as quartz), and contains larger amounts of iron, magnesium, and calcium than the continental crust, which is about 60 percent silicon dioxide.

Below the crust, extending approximately another 1,800 miles, is the mantle. The mantle has an average composition of about 45 percent silicon dioxide and 38 percent magnesium oxide, with the remainder of its bulk composed of various iron, aluminum, and calcium compounds. The mantle is divided into several zones or layers with different physical and chemical properties. The uppermost layer, lying immediately below the crust, is relatively rigid. This layer is thinner than thirty miles under some ocean basins, but under older continental areas it can be over 120 miles thick. Together this uppermost layer of the mantle and the crust attached to it is known as the lithosphere, which means "rocky shell" in Greek and refers both to its rigidity and to the ordinary rock of the crust that constitutes its upper surface.

Under the lithosphere lies the relatively weak and soft asthenosphere (literally, weak or glassy sphere). Below the asthenosphere the mantle becomes more rigid again.

Beneath the mantle lies the core, which, like the mantle, has layers. The outer core, extending from a depth of about 1,800 to 3,200 miles, consists of a liquid that is an alloy of mostly iron with some nickel. The inner core, extending from the edge of the outer core to Earth's center (about 3,960 miles deep), is composed of a solid iron-nickel alloy.

If we take the metaphor of the Earth-onion further and launch it into space to orbit the sun, we would notice that all the layers spin together in the daily rotation about the axis. Being closer to the axis, the inner layers spin more slowly than the outer, but in the course of any given period of time, each layer would make the same number of revolutions. This is not true for Earth, however.

Recent work by seismologists Xiaodong Song and Paul Richards of the Lamont-Doherty Earth Observatory in Palisades, New York, suggests that the inner core of the planet rotates faster than the rest of the planet. According to their research, which rests on changes over time in shock waves caused by earthquakes bouncing off Earth's interior, the inner core turns about one degree farther than the planet as a whole each year. In approximately 360 years, the inner core, rotating on its own, would make one more complete revolution than all the layers outside it. The faster rotation of the inner core may explain Earth's magnetism, a still poorly understood phenomenon that Albert Einstein once declared one of the great unsolved problems of modern physics.

According to a computer model developed by Gary Glatzmaier and Paul Roberts, the inner core is solid iron spinning in an electrically conducting liquid outer core. The result is a kind of dynamo that produces an electrical charge that in turn maintains a magnetic field. Glatzmaier and Roberts, working independently from Song and Richards, also posit an inner core that rotates separately from the rest of the planetary layers at about the same rate of one degree more per year. If the speed of rotation should change, the magnetic field would shift with it—a possible explanation for why Earth's magnetic poles swap position every few hundred thousand years or so.

Continents on the Move

The interior of Earth is an intensely hot place of seething and churning, where rocks slide and move around one another, even melt into liquid. At breaks and cracks in the crust, molten rock erupts onto the surface, adding to the crust. In other areas ancient rock is dragged back toward the center of the planet, to be reheated, melted, and rejuvenated. The vast majority of

Earth's surface is relatively young rock. In comparison, its close neighbors—the moon, Mercury, Mars, and, as far as we can tell, Venus—have been largely inactive for billions of years, and the rock that covers their surfaces is correspondingly ancient. Earth is younger, and it is still on the move. That makes all the difference in how our planet works, a fact we have begun to appreciate only in this century.

As we discussed in regard to the nonexistent continent of Lemuria, scientists of the nineteenth and earlier centuries assumed that the major continental landmasses were fixed in place. Land bridges might arise to connect continents from time to time, but the continents themselves remained resolutely where they always had been.

Still, as practically every observant elementary schoolchild has noted from looking at a world map or a globe, the shapes of the continents seem to fit together nicely, something like the pieces of a global jigsaw puzzle. The east coasts of North and South America slip neatly into the west coasts of Europe and Africa. This basic observation, combined with a wealth of other data—such as similarities between the animal species of Europe and North America—led the German scientist Alfred Wegener (1880–1930) to propose, in about 1912, that the continents had once been joined together and since had slowly drifted apart.

For decades the concept of continental drift was taken with little seriousness by most geologists, especially in the United States. Essentially the scenario seemed impossible: How could huge continents float or drift apart from each other? Wegener offered no explanation for the drift; he said only that it happened. Without a mechanism to explain it, the theory fell short.

However, a variation of Wegener's theory known as plate tectonics has transformed geology since the 1960s and today is accepted by virtually all earth scientists. Plate tectonics forms the unifying theory for most of the geological structures observed on the surface of the planet. Plate tectonics not only moves continents, but also raises mountains, creates new sea floor, destroys and recycles old sea floor, and causes volcanoes and earthquakes. It is plate tectonics that distinguishes our living planet from the dead planets of Mercury, Venus, and Mars.

Earth's lithosphere is divided into about eight major plates and numerous smaller ones, all of which are continuously moving. Where they meet, they slip past each other or collide. Sometimes they pull apart, opening a gap through which the liquid mantle wells up from underneath. Essentially the continents are carried on the tops of the plates, like children at play riding piggyback on their fathers. The continents are less drifting than going along for the ride as the lithospheric plates they are attached to move. The plates do not travel very fast from a human perspective, typically from less than an inch to twelve inches a year, but this is fast enough to have caused very major changes over the past few hundred million years.

Exactly what causes the plates to move has been a topic of heated debate and discussion for some time. Today most scientists believe that convection currents in the molten mantle propel the lithospheric plates on their slow but inevitable paths. As you can easily observe in a fresh cup of coffee or a pot of cooking soup, hot liquid rises to the surface, flows along the surface for some distance, and cools at the surface. The cooler liquid sinks below the surface, only to be heated up and start the cycle once again. This movement of liquid from hot to cool to hot again forms a convection cell. It is believed that giant convection cells driven by heat flowing continuously out of the center of Earth exist in the liquid mantle. As they cycle, the flowing mantle pulls the overlying, rigid lithospheric plates along.

The movement of two lithospheric plates away from each other leaves a gap in the solid crust. Initially the rock in this area may collapse, and form what is called a rift valley, but as the plates move apart, hot molten rock wells up to fill the void. This material originates from the mantle and forms new oceanic crust; the process is often referred to as *sea-floor spreading*. The classic example of two plates moving apart from each other is the Mid-Atlantic Ridge, which runs roughly north-south through the Atlantic Ocean.

In some places two lithospheric plates may converge or collide. As they push into each other, one of several things can happen. If the colliding edges of the two plates both bear oceanic crust, either plate may be forced, or subducted, under the other plate. A deep trench forms on the surface where the plates warp downward. As the subducted plate moves down into the hot mantle, it melts, and the lighter rock components rise to the surface, forming volcanoes. New continental crustal material, formed from the recycled oceanic crust, collects behind the trench. The northern and western edges of the Pacific Ocean, sometimes known as the Ring of Fire, are lined with convergent plate boundaries and, as a result, with a multitude of volcanoes.

If the leading edge of one colliding plate contains continental crust and the leading edge of the other plate contains oceanic crust, then the plate with oceanic crust will always be subducted under the plate-bearing continental crust. This occurs because the oceanic crust, which is composed of basalt, is thinner but denser than the continental crust, which is principally granite. Essentially, continental crust weighs so little compared to basaltic crust and mantle material that it cannot sink or be subducted to any significant depth. Instead, it "floats" on top. An example of this type of subduction is seen along the western coast of South America.

If the leading edges of two colliding lithospheric plates both contain continents, neither can be subducted because they are both so relatively light. Instead, the continents crash into one another, crumple, and deform, often raising imposing mountain ranges. The lofty Himalayas are the result of two continental landmasses crashing into one another. Eventually the plates will lock together and motion between them will stop.

Finally, two plates may slide past one another; their edges grind and slip along what are commonly termed transform faults. This is the least dramatic type of plate motion because it does not usually involve either volcanoes or mountain building. Still, it can be literally earthshaking, as the residents of California are periodically reminded. That state's many transform faults mark the northward sliding of the Pacific plate against the North American plate. The rocks on the surface do not flow past one another slowly and smoothly. Rather, they hang up like gears without grease, building pressure that is released suddenly in the sharp jerks and jumps of earthquakes.

The scientific beauty and elegance of plate tectonics is that it explains so much of the scientific data that geologists once found fundamentally unexplainable. It provides, for example, an excellent way of understanding the interlocking fit of continents, the geological function and geographical concentration of volcanoes, the geophysics of earthquake zones, and similarities in the animal forms on distant continents (such as moose and brown bears in Europe and North America, monkeys in Africa and South America).

The other advantage of plate tectonics is that it fits into the generally uniformitarian frame of mind most earth and life scientists bring to their work. Plate tectonics says that the physical world is changing, but only very slowly. After all, even at a blistering plate-tectonic rate of twelve inches a year, it would take over five millennia for a continent to move a mile, over 5.25 million years for it to cover 1,000 miles. The theory provides a congenial variety of philosophical comfort, offering a model of apparent and general similarity altered only by gradual, incremental change lacking drama, abruptness, or suddenness.

Which is precisely the problem. Plate tectonics delivers an excellent theory for understanding a great deal about our planet, but it cannot explain sudden, rapid, even catastrophic changes, like that sudden warming of the climate revealed in the Greenland ice cores from circa 9645 B.C. Plate tectonics has nothing to say about such a quick alteration in surface conditions on Earth. It is an excellent theory for what it explains, yet it falls short of accounting for it all. Other ideas are needed.

Hapgood: Pole Shift

Some of those ideas come from Charles Hapgood, whose work on the Piri Reis map was discussed in chapter 4. Hapgood was drawn to the study of ancient charts because of his interest in the possibility of relatively sudden large-scale movements in Earth's surface. He published a book on the topic, called *Earth's Shifting Crust,* in 1958, even before his book on ancient maps appeared. As the plate tectonics revolution of the 1960s reshaped geology, Hapgood stayed abreast of the new ideas and eventually revised his original book under a different title, *The Path of the Pole.*

Hapgood deserves credit on at least two grounds. For one, he was always a serious scholar, a man of remarkable intellectual integrity. Though trained as a historian rather than as a scientist, he came to his work with dispassion and discipline, sifting the facts as he saw them in the search for a workable explanation. His second accomplishment was rescuing the idea of pole shifting from earlier writers like Immanuel Velikovsky and Hugh Auchincloss Brown, who were more visionaries in touch with their private muses than scientists on the trail of an hypothesis. Brown, for example, argued that the entire planet suddenly shifted its axis of rotation, an argument he based on both the existence of frozen mammoths in Siberia, the biblical story of the flood, and the geological age of Niagara Falls. Similarly, Velikovsky used his own reading of mythology to argue that in the middle of the second millennium B.C., Jupiter had catapulted a piece of itself into space. The close passage of this chunk of planetary material, which would later settle down and become Venus, unleashed unspeakable cataclysms on Earth, including a shift in the axis of rotation. Hapgood freed the idea of polar shifting from such blather and, as we shall see, may have set geological thinking into a new and productive pathway.

As we all know, Earth spins or rotates, and the period of its rotation is the way we define the day. If, while you are carving Earth open to examine its layers, you draw a pencil line to mark the axis on which the planet spins, the line would emerge through the planet's surface at the north and south poles. In essence, the spin axis remains fixed in place over the short term relative to the plane of Earth's orbit around the sun and other astronomical markers, while the rest of the planet spins about it. Hapgood argued that the poles marking this fixity haven't always been where they are now.

According to Hapgood, the lithosphere of Earth has slipped over the inner layers and moved the poles at least 200 times in the past 600 million years. In just the past 80,000 years, the poles have moved substantially no less than three times. At the beginning of that period, the North Pole stood in what is now the Yukon Territory of Canada. From there it shifted eastward, reaching the Greenland Sea at approximately 73 degrees north latitude and 10 degrees east longitude, then slipped back to the west, reaching Hudson Bay at 60 degrees north latitude and 83 degrees west longitude by 50,000 years ago. It remained there until about 17,000 or 18,000 years in the past, when it moved at a rate of approximately 1,000 feet a year toward its present position, well north of Hudson Bay, above the many islands of the Canadian High Arctic. This movement was complete, Hapgood says, by about 12,000 years ago, or circa 10,000 B.C. On average, each pole shift has covered a distance of about 30 degrees of latitude, or 2,000 miles, approximately one-third of the distance from the equator to the pole, and it has taken several thousand years to reach completion.

Hapgood based his analysis in part on what is called *paleomagnetism.* When volcanic rocks cool from their molten state, they preserve an imprint of the Earth's magnetic field at the time they were formed. The magnetic pole isn't exactly the same as the geographical pole defined by the spin axis, but the two are close enough that we can use a compass to find north. Like a compass, the weak fossil magnetism remaining in volcanic rocks points to ancient north. Additionally, the angle of the magnetism in the rock indicates the latitude at which it was formed. It lies flat to the horizon at the equator, but dips straight down at the pole, assuming intermediate angles at points in between. Thus the magnetism in ancient volcanic rocks in a given location can provide data on whether the pole was located in the same place when they cooled as it is now, and whether the rocks have remained at the same latitude or been on the move since their formation.

In Hapgood's view, the paleomagnetic data make sense only if one theorizes that periodically the entire crust of the planet slips over the mantle. The planet continues to rotate as it always has, but after the crust slips, different land is situated at or near the poles than beforehand. The effects can be both catastrophic and creative, as formerly arctic landmasses move closer to the tropics and once-tropical realms are borne north into temperate and even arctic regions. As Hapgood sees it, the frozen mammoths of Siberia show just how profound an effect such a shift can have on living beings—including, presumably, the humans who have been spreading out over Earth's surface during the last three substantial pole shifts.

Despite Hapgood's resolutely meticulous mode of work, little of the data he cites has stood the test of time. Take the frozen mammoths as an example, a bit of evidence that has been trumpeted in a number of places, including the Charlton Heston–narrated TV documentary *The Mysterious Origins of Man.* As the oft-repeated story goes, numerous mammoths have been found in Siberia in such a perfectly preserved state that they must have been flash-frozen. In some cases, the story continues, the thawed meat was fresh enough to eat, and the stomachs of the animals contained undigested flowers, clear evidence that they died in the summer and that their deaths lasted only seconds at most.

Give this story any scrutiny, however, and it doesn't hold up.

The number of mammoths found frozen is but a few, and the meat, although edible, was nothing you'd try to eat unless you were a vulture. In some of the mammoths, the meat had begun to putrefy even before it froze, indicating that death came well before freezing. As for flowers among the stomach contents, recently ingested plants are so fundamentally tough that they often survive the decay of an animal after death. Rather than invoking crustal slippage or pole shifts to explain the few known frozen mammoths, it is more reasonable to look for other explanations. The animals may have been caught in blizzards, drowned in an icy stream or lake, smothered in a landslide or avalanche, or trapped after a stumble into a glacial crevasse.

More to the geological point of Hapgood's argument, however, are the paleomagnetic data cited to prove the shifting path of the poles. In the decades since *The Path of the Pole* was published, new and better data have been gathered that undercut Hapgood's argument. The poles may well have shifted, but nowhere near as far nor as often as Hapgood argues.

Finally, the most glaring omission in Hapgood's argument is his inability to come up with a mechanism for crustal shifting. He describes what happens, but he can't say why. He himself saw the problem and spelled it out. "It is necessary to admit, in the first place, that at the present time there is no satisfactory explanation of the *modus operandi* of displacements in the lithosphere," Hapgood wrote.

Yet, as we shall see, Hapgood may have been on the right track. He used bad data and poor science, and he certainly got the details wrong. Still, his work likely pointed in the right direction.

Noone: The Dangerous Planets

Hapgood's ideas have made their way into the writing of a number of other authors, including Rand and Rose Flem-Ath, whose work on Atlantis appeared in chapter 4, and Richard W. Noone, who wrote the successful *5/5/2000: Ice: The Ultimate Disaster.*

Noone's work is of interest for two reasons. Unlike Hapgood, who sees crustal shifts as requiring thousands of years—which is much faster than plate tectonics—Noone maintains they can happen practically overnight, with predictably disastrous consequences. And, again unlike Hapgood, Noone offers a theoretical mechanism to explain just how Earth's surface can undergo such rapid and profound change.

Like Hapgood, Noone is not a scientist—when he's not writing, he runs a nonhazardous chemical business—but rather an amateur who brings to his work an extraordinary enthusiasm for ancient civilization and a powerful interest in scenarios of catastrophe. Sometimes these proclivities collide, producing a book whose subject matter careens from the Egyptian pyramids to the torture of Knights Templar in fourteenth-century France to the toothpaste-like consistency of rock under polar glaciers.

Still, the basic thrust of Noone's argument deserves attention. It begins with the evidence that Earth was once home to a widespread, highly sophisticated civilization whose remnants we see in the Mayan and Aztec ruins, the pyramids of Giza, and the esoteric beliefs of the Freemasons and the Rosicrucians. This old world was destroyed abruptly in a sudden, massive movement of the crust that utterly altered Earth's climate and caused catastrophic earthquakes, tsunamis, and volcanic eruptions, all of them on a scale we cannot imagine. The "ice" in Noone's title refers to his theory that under the right circumstances the north polar ice cap can slip and slide from its present position, hurrying southward like a great solid tidal wave, destroying everything in its path.

Noone argues that this kind of catastrophe, which has happened before, will happen again soon, on May 5, 2000. The effects will be devastating. The polar ice caps will lurch toward the equator with a sudden terrible roar. The moving ice will raise a tidal wave in the North Atlantic that will crash into New York at a height of 300 feet. On that fateful day, bid good-bye to Broadway.

Noone is willing to be so specific about the date of this imminent destruction because he maintains he understands the mechanism that drives Hapgood's crustal shifts. On the day of his prediction, twenty-four hours after the appearance of a new moon, the planets Mercury, Venus, Jupiter, and Saturn will line up. The gravitational pull of this unusual planetary arrangement will tug so fiercely on Earth's crust that a shift in the surface layer, and most prominently the polar ice, is practically sure to result.

On first examination, Noone's idea sounds interesting as well as terrifying. As we have understood ever since the time of Isaac Newton, gravity is a powerful universal force, one that can reach across tremendous distances of space to affect the behavior of astronomical bodies. Even better, Noone's idea is one that can be examined logically to determine its scientific worth.

And that's where it falters seriously. To begin with, the planetary alignment of May 5, 2000, isn't really all that much of a line. The planets will be loosely gathered at the time, but they will be on the far side of the sun from Earth. In the same general direction, yes, but not lined up, and with the vastly greater—and normal—gravitational pull of the sun between them and Earth.

Also, although planetary alignments of this sort are unusual, they are hardly rare. Using a planetarium software called Voyager II, Brian Monson of the Physics Department at the University of Tulsa calculated that in the past thousand years there have been ten planetary alignments, happening on average once a century. One occurred as recently as 1982, a year during which, as we all know, the world did not end. On September 2, 1861, the planets were grouped much more closely than they will be on May 5, 2000, and, again as we all know, the world did not end.

Nor is it likely to do so soon. Earth is 4.5 billion years old. During that long period, planetary alignments at the rate of one every hundred years have happened approximately 45 million times. Several dozen of those alignments have occurred since humans learned to write and were studying the skies with great intensity (remember the sky bulls of Lascaux, in circa 15,000

B.C.). Had a sudden disaster fallen at the same time as a planetary alignment, surely somebody would have noticed and recorded it.

In fact, nothing about the physics of gravity should lead anyone to lose sleep over a planetary alignment. The fact is that the other planets, aligned or unaligned, have little gravitational effect on Earth.

One planet can have two types of gravitational effect on another planet. One is simple gravity, that business of the apple falling from the tree onto Newton's ever-thinking head. Simple gravity describes another planet's pull on Earth, in the same way Earth pulled on the apple that hit Newton. The second type of gravitational effect is tidal force, a more complicated phenomenon that effectively works like a stretching force on our planet. This stretching on the world's oceans gives us tides.

If gravity were strong enough, it could pull Earth out of its current orbit and into some new and certainly disastrous path. A tidal force of sufficient strength could stretch the planet until it tore in half, with equally calamitous consequences. But don't worry. Neither simple gravity nor tidal force, even in the case of a planetary alignment, is anywhere near powerful enough to do harm to Earth.

Our safety comes from basic physics. Two factors affect gravity: the mass of the other object and its distance. Basically, the more massive and closer something is, the greater the gravitational pull it exerts. Gravity increases linearly with mass—that is, an object twice as massive as another has twice the gravitational attraction—but it decreases with the square of the distance. Assume that an object of a certain mass is a certain distance away. Now you move the object twice the distance away. Since gravitation declines by the square of the object's distance, its pull falls not to half of what it was but to one-fourth ($1/2^2 = 1/4$). Move it ten times farther away, and the gravitational pull decreases to one one-hundredth of its former value ($1/10^2 = 1/100$). Obviously, gravitation falls very fast with distance.

Tidal force is much like gravity, except that its force falls off with the cube of the distance. Move that object of a certain mass twice as far away, and its tidal effect drops to one-eighth ($1/2^3 = 1/8$). At ten times the distance it falls to one one-thousandth ($1/10^3 = 1/1{,}000$). Distance has an even more profound limiting effect on tidal force than on gravity.

If we know the mass and the distance of an object, calculating its gravitational influence and tidal pull on Earth is straightforward. The moon is the hands-down winner among all the planetary bodies. Although it is only one-eightieth the mass of Earth, it is so close that it exerts the largest gravitational and tidal effects on our planet. The nearest planet, Venus, is about two-thirds as massive as Earth, and about sixty-five times more massive than the moon, but it is 150 times farther away than the moon even at its closest approach. What with squaring the distance for gravitation and cubing it for tides, Venus's gravitational effect is 0.006 that of the moon and its tidal effects only 0.00005 as large. Even giant Jupiter, which is more than 300 times as massive as Earth, has just 1 percent as much gravitational effect and 0.0006 percent as great a tidal influence as the moon.

Add all the planets together, and, even at their closest approach to Earth, they exert less than 2 percent of the moon's gravitational pull. As for total planetary tidal pull, it's a mere 0.0058 percent that of the moon's, even at closest approach.

These numbers will drop even further during the planetary alignment of May 5, 2000. The reason, again, is distance. Earth will be on one side of the sun, while the grouped planets will be on the opposite side. That arrangement adds some 185 million miles to the distances between Earth and each of the other planets and relegates their gravitational and tidal influences to insignificance. For example, on May 5, 2000, Venus's tidal influence will be only one five-hundredth what it is at closest approach to our planet.

Let's assume that Noone merely has the date wrong. What if, at some other point in the future, all the planets lined up at their closest approach to Earth. Would the 2 percent increase in gravitational influence that would result from this alignment have a major effect on our planet?

The answer is no. In fact, Earth is subject to much greater gravitational variation all the time, without harm. The moon's orbit is an ellipse, not a circle. At its closest approach (perigee) the moon is slightly over 26,000 miles nearer Earth than at its farthest orbital point (apogee). Given the powerful effect of distance on gravitation, the moon's orbital variation varies its pull on Earth by approximately 25 percent between closest and farthest points. This means that every two weeks Earth is subjected to a gravitational fluctuation over ten times greater than the pull exerted by all the planets lined up at their closest point to us. And nothing catastrophic happens.

Simply put, the planetary alignment of May 5, 2000, bodes neither ill nor portent. Still, if one looks past Noone's need for a sudden, dramatic event, he is making a valid point: A shift in Earth's axis would have a profound effect on the planet.

Strain: Small Shift, Big Effect

Just how big that effect would be has been described in meticulous detail by Mac B. Strain, in his book *The Earth's Shifting Axis.* A professional civil engineer now retired from the National Mapping Division of the U.S. Geological Survey in Denver, Strain traces out the remarkable changes in Earth's surface that would be caused by even a small, slow shift in the axis. Here we are talking about an actual shift in Earth's spin axis that involves the movement of the entire mass of the Earth—not just the outer crust or shifting relative to the vast inner mass of Earth as posited by Wegener, Hapgood, or plate tectonic theory.

As we saw in chapter 3, Earth may be a perfectly round sphere in a schoolroom globe, but in fact it is an oblate spheroid, thicker at the equator than at the poles. The planet's spin creates a centrifugal force that lifts the faster-spinning equator away from the slower-spinning higher latitudes. Thus a radius drawn from the center of Earth is 13.5 miles longer at the equator than at either pole. The resulting planetary shape is what we geologists call the *geoid,* which represents the shape of Earth as defined by sea level.

The geoid's bulging shape has a profound effect on altitude. If the wrinkled bump on the surface of Earth that represents the Himalayan area surrounding Mount Everest, which lies at approximately 28 degrees north latitude and protrudes 29,028 feet above sea level, were moved south to the equator, the peak would be only 13,434 feet tall—a good weekend climb, but nothing to have challenged the likes of Sir Edmund Hillary and Tenzing Norgay, the first men to reach the summit of the world's highest mountain. Why the change in altitude? The geoid surface at the equator is over fifteen thousand feet farther from the center of Earth than it is at 28 degrees north latitude. As a result, the Mount Everest bump that stands out so tall at 28 degrees north latitude is much diminished in relative height when it moves to the equator. Given the same treatment, Mount Denali in Alaska would lie a remarkable 35,958 feet below sea level. The shift works the other way as well. If the Challenger Deep in the deep-sea Mariana Trench—the lowest point on the crust at 35,810 feet below sea level, located at approximately 15 degrees north latitude—were at the pole, it would lie 31,946 feet above sea level. Because it is a depression, the Deep would still be under water, but it would take the form of a lake, not an ocean.

The most dramatic change in Earth's surface caused by a shift in the axis is the moving boundary between ocean and dry land. Assume for a moment that, all other things being equal, the axis shifts a full 90 degrees. Assume, too, that the solid portion of Earth remains rigid and does not change shape. This shift places the new poles on what is now the equator, and moves the former poles to the new equator. Because of the centrifugal force of the Earth's spin and the combined gravitational and tidal pull of the moon and the sun, ocean waters pour over the old poles to a depth of 13.5 miles, while the seas drain away from the portions of the former equator now at the poles, elevating the area an equivalent 13.5 miles, shallowing the ocean and even creating dry land in some places.

In reality, however, Earth is not rigid, but somewhat plastic and malleable. The solid Earth would deform as the axis shifted, yet dramatic changes—nearly impossible to predict in their particulars—would certainly result in the event of a large-scale axis shift.

Yet, according to Strain's calculations, even a much smaller shift has a marked effect on the distribution of land and sea across the surface of the globe. If the axis shifted only one degree from its north polar current position of 90° north latitude to 89° north latitude and 70° west longitude—a move of about seventy miles—the change in the geoid would raise sea level in some areas and drain it in others. The effect would be most pronounced along the 70° west and 110° east longitude lines at the midway points between the poles and the equator—namely 45° north latitude for the northern hemisphere and 45° south latitude for the southern. Some spots would gain more than 1,200 feet, mirroring an equivalent loss in others.

A look at the new shape of the northern hemisphere under the conditions of such a scenario gives us a sense of how much the configuration of land and sea could change with only a one-degree axis shift to 89° north latitude and 70° west longitude. The United States, except for Alaska west of 160° west longitude—the approximate position of Wainwright on the Arctic Ocean and Port Moller on the Alaskan Peninsula separating the Bering Sea from the Pacific Ocean—lose sea level as the ocean drains away to give the geoid its new shape. The Pacific Coast looks pretty much the same as it does now, because the continental shelf is so narrow, but Canada's shallow Inland Passage, Puget Sound, and San Francisco Bay empty. Lower California becomes twice as wide as it is now, and Central America widens, owing to the exposed continental shelves on both east and west coasts. The shorelines of Honduras and Nicaragua, for example, extend 100 to 200 miles east of their current locations, and the Yucatán Peninsula reaches an additional 140 miles north. You'll have to travel a similar extra distance to reach the waters of the Gulf of Mexico in Texas and Louisiana. Florida more than doubles in width and extends as solid land through what are now the separate islands of the Florida Keys. The east coast from Florida to New Jersey moves out into the Atlantic some seventy miles. A new section of land about the size of the current state of Pennsylvania rises up off Cape Cod as the ocean spills off what used to be the Georges Banks. An even larger section of land is appended to Nova Scotia though the exposure of the Sable Island Bank, and the draining of the Grand Banks moves the Newfoundland shoreline 300 miles east. Most of Hudson Bay also drains away, and many of the islands of the Canadian High Arctic are joined to the mainland by new land bridges.

Powerful changes occur, as well, in the center of the country. Lake Michigan, which is almost thirty feet higher at Chicago and swamps much of that lively windy city, does again what it did in the last ice age. It flows not north through Lake Huron and the other Great Lakes to the Saint Lawrence River and ultimately the Atlantic, but south into the Des Plaines, Illinois, and Mississippi rivers toward the Gulf of Mexico.

Like North America, most of Western Europe loses water. The North Atlantic islands gain size, so much so that Britain and Ireland connect to the continent. Now you can drive from Dublin or London to Paris without taking a ferry or using the Chunnel.

By contrast, the geoid's reshaping in Eastern Europe and Asia causes massive flooding. Istanbul lies 232 feet lower than it does now, putting it under water and turning the narrow channel of the Bosporus into a wide reach of the Mediterranean. The Black Sea expands and may well connect with the Caspian Sea, perhaps even the Aral Sea, extending the Mediterranean system deep into

Central Asia. The flooding hits a maximum along the 110° east longitudinal meridian, where seawater floods the lower reaches of the Indus, Ganges, Mekong, Yangtze, Yellow, and Amur rivers. Coastal cities like Karachi, Bombay, Singapore, Rangoon, and Shanghai disappear, and extensive reaches of low-lying land, including the whole of Bangladesh and many of the best rice-growing regions of southern Asia, slip under the waves. In the far north, low-lying tundra and taiga, such as the West Siberian Plain, are likewise swamped out by the invading Kara, Laptev, and East Siberian seas. In the North Pacific, much of the western end of the Aleutian island chain shrinks or disappears. The Bering Strait grows larger, allowing warmer Pacific currents to flow northward and fundamentally altering the patterns of weather generated in the Bering, Chukchi, and Beaufort seas.

Of course, this entire scenario is fictional, an imagined set of consequences flowing from a hypothetical event. It is based, too, on a number of assumptions concerning the rigidity of the Earth's solid surface, and the behavior of the mantle and lithosphere in reaction to the new centrifugal forces acting upon them if the axis shifted. These assumptions may not be realistic, but nobody knows for certain what the consequences of an axis shift would be.

Strain's calculations raise important questions. Has such a shift ever occurred? If it has, what effect did it have?

Kirschvink: True Polar Wander and the Biological Big Bang

The flip side of a massive extinction like the Permian or the Cretaceous is an explosion of new life-forms. Once in a while in the geological history of Earth this has happened, with a suddenness and a creativity to take away breath and inspire awe. The record of the largest such explosion resides in a Canadian rock formation known as the Burgess Shale.

Located in British Columbia, the Burgess Shale is particularly important because it contains the fossils of many soft-bodied creatures from over 500 million years ago, during the period known as the Cambrian. Soft-bodied animals, unlike those with bones and shells, leave fossils only under rare and unusual circumstances, and the Burgess Shale provides an extraordinary and precious look at a remarkable biological event.

At the start of the Cambrian period, most life-forms were single-celled. The few multicelled animals then in existence left no more fossil evidence than trails, burrows, and a few body impressions that are hard to connect with any contemporary species. Then, in a remarkable 10-million-year period stretching from approximately 540 to 530 million years ago, the ancestral forms of virtually all modern groups of animals appeared, including the first chordates, the group that gave rise to vertebrates and eventually to us humans.

This amazing burst of life is commonly called the Cambrian explosion, and is sometimes referred to as "the biological Big Bang." Without the extraordinary events of the Cambrian period, you and I might never have come into existence to share this book.

Equally remarkable, the diversity of species arising in the Cambrian was stunning. The creatures found in the Burgess Shale exceed in anatomical variation the whole range of invertebrate animals inhabiting present-day Earth. Today, for example, we know of about 1 million species of Arthropoda, a phylum that includes such animals as insects, crustaceans, spiders and mites, centipedes and millipedes, and horseshoe crabs. The Burgess Shale reveals fossils from all of these groups, plus a number of other apparent arthropods that are distinct types of animals with body plans unknown today. In the case of arthropods, the Cambrian world boasted more diversity than ours does.

Usually we think of evolution as starting with one animal or plant form and producing new forms from it, like small branches radiating from the trunk of a tree. The Cambrian explosion shows the opposite. An almost unimaginable burst of creativity led to an enormous diversity of life-forms that was then winnowed and narrowed over the eons to produce the species we have now.

Fully understanding how this process occurred could fundamentally alter the way we view evolution.

The Cambrian explosion raises another important question: Why? Before that explosive 10 million years, life was doing just fine as it was. What change happened to prompt a burst of evolutionary fervor the likes of which Earth has never seen again?

A fascinating hypothesis to explain the Cambrian explosion has been proposed by California Institute of Technology geologist Joseph L. Kirschvink and his colleagues Robert L. Ripperdan of the University of Puerto Rico and David A. Evans, also of Cal Tech. Kirschvink proposes that the entire crust and mantle slipped a full 90 degrees over the core during the Cambrian, forced landmasses into new relationships, and fundamentally altered the conditions for life on Earth. The poles stayed where they had been. Everything else changed.

Like Hapgood's work, the Kirschvink model is based on paleomagnetic data, which now are much more reliable than when Hapgood studied them. Combining the paleomagnetic information with other data that allow the dating of rocks gave Kirschvink and his colleagues the ability to track the movements of continents over hundreds of millions of years.

The research began in Australia, which apparently made some remarkable movements during the Cambrian explosion. At the time, Australia was part of a larger landmass called Gondwana, which also included South America, Africa, and India. From 600 million to 540 million years ago, Australia remained in place at the equator. Then it began to move, as did the whole of Gondwana, starting approximately 535 million years in the past. Over the next 15 million years or so, Australia remained at the equator, but swung counterclockwise in a half-circle, completing a 90-degree turn. Gondwana made the same twisting movement, so that its landmass, which had been divided more or less equally between the northern and southern hemispheres, lay now almost entirely in the southern hemisphere.

To validate this motion and check the data, Kirschvink compared the Australian geomagnetic data with reliably dated rocks from North America. When Australia and Gondwana began moving counterclockwise, North America lay close to the South Pole as a separate landmass divided from Gondwana by the long-gone Iapetus Ocean. As the twist occurred, North America moved north until it came to rest on the equator. What had been a polar territory was now tropical.

This didn't happen overnight. Rather, the process took about 15 million years. Still, such a huge global swing means that the continents traveled several feet a year, a rate many times faster than the movement of plate tectonics. Kirschvink sees it as a different process, one called "true polar wander." For at least the past fifty years, geophysicists have thought such motion was theoretically possible. If Kirschvink is right, it has actually happened.

Unlike plate tectonics, which is driven by convection currents in the mantle, true polar wander comes from imbalances in the planet's distribution of mass in the crust and mantle. During true polar wander, the entire solid part of Earth moves together as one mass over the liquid core. Earth maintains its original spin axis, but the landmasses are rearranged in relation to it.

The physics of this process can be modeled by gluing lead weights to opposite points on a basketball. If you set the ball on a slick gym floor with the weights at its midline—in planetary terms, the equator—and spin it, it will turn round and round with the weights remaining on the equator. Spin the ball on the weight—with one heavy point at one pole and the other heavy point at the other pole—and the ball will follow a path that in time moves the weights to the equator. As the laws of physics require, the ball has aligned its maximum moment of inertia—represented by the weights—with its spin axis as marked by the poles.

Fundamentally the same thing happened to Earth during the Cambrian period. The difference is that, rather than the whole ball shifting, Earth's surface layers moved with respect to the core. The spin axis remained where it was; the landmasses sorted themselves out in a new pattern to bring the maximum moment of inertia into alignment.

The imbalance arose, according to Kirschvink and his colleagues, about 550 million years ago, as violent earthquakes tore continents apart, pushed them together again, and raised huge volcanic areas at the continental seams. Some large and as yet unknown event brought bits of many continents together as a large section of sea floor sank and a new volcanic range rose near the South Pole. This new slab of rock in the polar position, with most of the planet's continents crowded together in one imbalanced landmass, was enough to trigger true polar wander.

Data on ocean currents tend to support Kirschvink's model. With the landmasses on the move, circulation in the seas was constantly upended. Every change in oceanic currents released carbon deposits buried in the deep waters of the seas, causing an upswing in dissolved carbon. At least a dozen such swings occurred during the Cambrian period, an anomaly that has long troubled scientists. True polar wander explains it.

Some astronomical evidence points to true polar wander on planetary bodies other than Earth. On the Martian equator is a gigantic volcano known as Tharsis. The largest known planetary-surface gravity anomaly in the solar system, Tharsis is very dense and massive, and therefore exerts a stronger gravitational pull than the rock that surrounds it. Tharsis may have formed in its present position, but more likely it arose elsewhere on the planet, and then, because of its extraordinary mass, moved via true polar wander, like the weight on the spinning basketball, to its current location. An opposite effect is thought to have occurred on the asteroid Vesta. Struck by some fast-moving space object that left a massive crater, the asteroid reoriented so that the impact crater, which has less mass, ended up at the south pole.

Of course, all by itself, true polar wander didn't cause the Cambrian explosion. Rather, it created both new demands on animal forms and novel opportunities for them. The massive changes resulting from all this shifting and moving would have divided existing communities of species into smaller, more inbred assemblies, conditions that amplify genetic differences and accelerate evolutionary changes. And since the change kept going on over those 15 million years, every new adaptation was presented with a continuing series of new challenges and even more opportunities for evolutionary development. True polar wander created the special conditions that set the stage for the Cambrian explosion.

The Impact Possibility

There's no reason to believe that true polar wander has occurred only once in Earth's history, in the Cambrian period. Nor is there reason to believe that the rapid appearance of a volcanic range near one of the poles is the only possible cause. The asteroid Vesta, after all, is thought to have shifted owing to the mass imbalance caused by cratering from a collision with some other piece of space rock or ice. Indeed, some brilliant theoretical work indicates that the impact of a comet or asteroid with Earth could, under the right conditions, lead to a shift in the poles. This hypothesis is distinct from Kirschvink's model, which has the Earth's solid layers moving around over the liquid core through millions of years, while the spin axis remains the same. Instead, in a true shift of the axis, like the one modeled by Strain, the whole planet changes position and assumes a new spin axis in much less time.

The argument, proposed by the Italian mathematician Flavio Barbiero, is an attempt to explain the sudden warming of the climate 11,600 years ago, as seen in the Greenland ice cores. Building on the work of Charles Hapgood, Barbiero believes that the poles shifted rapidly, changing positions far more quickly than the thousands of years Hapgood proposed.

The key to understanding Earth's behavior in the event of a major impact, Barbiero argues, is realizing that the planet acts like a giant gyroscope. If a toy gyroscope is struck with a disturbing torque, its spin axis shifts, even if only for an instant. The same thing could happen to Earth, which is not solid, like a child's plaything, but semifluid, with oceans outside and a liquid core inside the rigid lithosphere. The centrifugal force of Earth's rotation affects each of these layers and deforms or displaces them, as we have seen in Strain's analysis of what a pole shift can do. It is possible, according to Barbiero, that Earth could be struck by an object with a torque large enough to overcome the force creating the current axis of rotation and establish a new axis—with, of course, a simultaneous shift in the position of the poles. Since so much of Earth is covered with a layer of oceanic water, this water shifts toward the new equator with enough force to deform and reshape the planet's mantle. Once the mantle's shape changes, the new axis of rotation becomes established, perhaps in a matter of minutes, hours, or days.

According to Barbiero's calculations, a rocky asteroid 1,100 yards across—not very large by the standards of our solar system—striking at the right angle and in the same direction as the gravitational pull of the sun and moon, would be more than large enough to cause such a shift. The effects of the collision would be devastating, but the aftermath of the pole shift would overshadow the short-term effects of the impact explosion. As the poles shifted, some parts of the crust would move up, others down, releasing huge earthquakes and volcanic flows. Winds of hurricane force mixed with torrential rain would sweep the continents. Water would avalanche into plains and valleys, and sea level would fluctuate both wildly and widely, perhaps in the form of a tide several hundred feet high moving around the globe. The planet's core would be affected by turbulence, particularly at its boundary with the mantle, possibly shifting the magnetic field and even reversing the magnetic poles.

The long-term effect on climate would depend largely on the orientation of the new axis of rotation relative to the ecliptic. Right now, this stands at about 23.5 degrees, and the planet's tilt is the reason why the temperate zones have seasons. If the axis of rotation moved so that it stood nearly perpendicular to the ecliptic, say at an angle of 5 degrees or less, the seasons would vary little and vegetation could grow all year long, providing a steady supply of food to plant-eating animals and to the carnivores that feed on the grazers and browsers. With the poles always bitterly cold, ice would accumulate into massive glaciers and sea level would drop. Tropical species could move north, and northerly species could move south, producing a mix of species found nowhere today. Barbiero maintains that this was precisely the situation up until the end of the last ice age, when mammoths, rhinoceroses, and lions, whose modern descendants are found only in the tropics, cavorted on the same grasslands as reindeer, bison, and bears, species that flourish in today's temperate and boreal zones. Then something happened, the poles shifted, and the old world came to an end.

Turning to the Stars

Barbiero's model is frankly speculative, a hypothesis based on certain physical laws that creatively examines a series of what-ifs and develops a scenario arresting in its drama. It may be entirely wrong. Yet even if it is, it forces us to do something important: to look outside Earth for the causes of events on our planet.

Geology, the discipline in which I make my professional home, has this way of looking down. When something happens on Earth, we want to explain it by means of Earth-only events: continental drift, volcanic activity, plate tectonics, earthquakes, and now even true polar wander. Yet in all this catalog of Earth's simultaneously creative and destructive forces, we have yet to find one that accounts for the sudden catastrophe at the end of the Bronze Age, the point at which we began. Can it be that something else is going on here? Can it be that we must turn our attention where Barbiero points us—to the heavens?

6

Heaven's Rain of Rock and Ice

MANY OF THE ANCIENT HEBREW MYTHS RETOLD IN THE OLD Testament depict the sky as menacing and dangerous, the terrifying source from which God's burning wrath descends to punish the sinful inhabitants of Earth. Consider the fate of Sodom and Gomorrah. Angered by the multitudinous crimes of the two cities' citizens, God resolved to destroy them, with the singular exception of Lot, the only righteous man living there. As the sun rose on the appointed day of destruction, two angels hurried Lot and his family out of town, just in the nick of time. Barely had Lot made it into the nearby small town of Zoar when, according to chapter 19 of the book of Genesis, "Yahweh rained on Sodom and Gomorrah brimstone and fire.... He overthrew these towns and the whole plain, with all the inhabitants of the towns, and everything that grew there." Abraham, the great patriarch of the Hebrew people, went to have a look at the destruction for himself. "Looking toward Sodom and Gomorrah, and across all the plain, he saw the smoke rising from the land, like smoke from a furnace."

Word about this kind of thing got around, even in the days before telegraph, telephone, and the Internet. When Jonah the prophet—with a little help from the famous whale—made his way to Nineveh in Mesopotamia, a city so great that walking across it took three days' time, he told its citizens that God was planning to destroy them for their wickedness. This threat was so credible that the Ninevites immediately did penance to atone for their sins, and God chose to spare them the promised fate of destruction descending from the heavens.

Sometimes the Hebrews took fire in the skies as a sign of prophecy. Ezekiel, a prophet to the Hebrews during their captivity in Babylon in the sixth century B.C., first felt the presence of God in celestial flame. "I looked," the opening of the book of Ezekiel says. "A stormy wind blew from the north, a great cloud with light around it, a fire from which flashes of lightning darted, and in the center a sheen like bronze at the heart of the fire." Among other wonders in the light, Ezekiel saw a throne, and "high up on this throne was a being that looked like a man. I saw him shine like bronze, and close to and all around him from what seemed his loins upward was what looked like fire; and from what seemed like his loins downward I saw what looked like fire, and a light all around like a bow in the clouds on rainy days."

Fear and Power in the Sky

The Hebrews were hardly alone in their attitude toward the heavens. Many ancient peoples looked up with a reverence that mixed awe at the sky's power with fear of the terrible destruction it could deliver.

Ovid, the classical Roman poet born in the later years of the first century B.C., provides the most detailed and dramatic version of the myth of Phaethon. The son of the sun god Apollo and a mortal woman, Phaethon requested of his father the right to drive the chariot of the sun in its daily passage across the vault of the sky. Enjoined by an oath to grant whatever wish his son requested, Apollo could not say no. Yet he warned Phaethon of the difficult road through the sky, the headstrong willfulness of the horses pulling the chariot, the dangers posed by the celestial beasts of the zodiac. Still, Phaethon demanded his wish, only to discover very soon that his divine father had had a point. Unable to control the horses and frightened by the terrible beasts of the sky, Phaethon dropped the reins. Loosed from their restraints, the horses lurched out of their usual course and pulled the chariot close to Earth's surface. In seconds, mountains caught fire, prairies dried and cracked, forests and croplands burst into flame, cities and their peoples perished in the blast of heat. All the snow and ice melted off the highest peaks, even lofty Olympus, and the rivers, from the least mountain streams to the mighty Nile, boiled into steam and disappeared, leaving only dust and sand where moments before water had flowed. Unable to bear any more, Earth herself cried out in pain and agony. To save the goddess-planet, the god Jove hurled a thunderbolt at Phaethon and the chariot, smashing it into a shower of burning pieces. Dead Phaethon himself fell to Earth ablaze, appearing to those below like a star plummeting from heaven.

A similar story comes from the opposite side of the world, namely Central America. The Annals of Quauhtitlán recount that the world-age known as the Sun of Air, or Ehcatonatiuh, was marked by a furious wind that knocked down buildings, pulled trees up by their roots, and moved boulders across the surface of the land as easily as a breeze sailing leaves over a pond. During this age, the great god Quetzalcoatl arose to teach humans the ways of virtue and the arts of living. When his ideas failed to take hold, he departed toward the east, where he encountered the sun and was burned up. From his ashes rose birds of shining plumage, and his heart ascended into the sky as the Morning Star. Thus began the age known as the Sun of Fire, so named because it was predicted that, like Quetzalcoatl himself, it would end by fire.

The great Hindu epic, the *Mahabharata,* paints the sky danger as a horrific bird with one eye, one wing, and one leg that hovered in the night sky, screamed, and vomited blood. When this terrible fowl appeared, dust came down upon the Earth in a thick choking blanket, and rough winds blew unceasingly. Fireballs crashed into the land with great hissing sounds, and a deep darkness descended, so that travelers had to light their way with torches even at high noon. The mountains resounded with explosions that tipped peaks and hillsides into the flatlands.

The text of the *Mahabharata* tells us what time of the year these events happened and provides something of a clue to the type of phenomenon that was associated with stories like Sodom and Gomorrah, Phaethon, and Quetzalcoatl. The terrible bird made its appearance during the Indian month of Karttika, which covers the last half of October and the first half of November. This period, which stretches from about October 15 to November 15, is considered a time of great portent and even danger in many cultures the world over. The ancient Celts, for example, made November 1 a one-day month called Samhain that marked the beginning of the new year—a seemingly odd choice, given that the date has no solar importance. On this day the spirits of the underworld flooded up out of their subterranean darkness onto the surface of the Earth, where they sometimes grabbed unsuspecting mortals and pulled them down to death in the nether regions. To avoid this fate, people dressed up like ghosts, goblins, and other monsters so the underworld spirits couldn't tell the living from the dead, a custom we preserve today in Halloween costuming. Likewise, Mexicans build elaborate altars to departed relatives for the November 1 celebration known as the Day of the Dead, and the Roman Catholic Church marks October 31 as the Feast of All Souls and November 1 as that of All Saints. Mortality fills the religious air in this brief, foreboding season.

The *Mahabharata* provides even more information as to why this period is so thick with thoughts of death and dying. The terrible bird affected the constellation we know as the Pleiades. During the October 15–November 15 period in question, the meteor shower known as the Taurid stream arises from the portion of the sky near the Pleiades constellation. In Hindu mythology, the catastrophes brought about by the horrible fowl had something to do with the Taurid's wild display of shooting stars.

This association of meteors with destruction from the skies is heightened by the powerful religious power accorded to the remnants of cosmic visitation. Egyptian tombs have been found to contain meteorites. Among the various treasures of the burial chamber of Pharaoh Tutankhamen was a dagger fashioned from meteoritic iron. When a meteorite fell to Earth in Phrygia, a region of the Middle East, in 2000 B.C., it was worshiped there for years, then later transported to Rome, where, according to the Roman historian Livy, it was venerated for another five centuries. To this day, Tibetan shamans and lamas prize the special power of tektites, natural glass beads formed from sediments in the extreme heat of a meteor impact. The Hopewell Native American culture in what is now Ohio and neighboring parts of the eastern United States made ceremonial tools from iron-rich meteorites, while the Indians of the Chihuahua desert of Mexico built special shrines for meteorites, one of which weighed over 700 pounds and was wrapped in the mummy cloth used for human burials. An even larger black meteorite has lain for centuries at the center of the shrine known as the Kaaba, a perfectly rectangular stone hut now housed inside the Great Mosque of Mecca. When Islam's founder, Mohammed, came to Mecca to claim the Kaaba for his one true God, the pagan idols fled, running away on stone legs. Only the black rock, reputed to have been a gift from the archangel Gabriel to the Hebrew patriarch Abraham, remained. Even today it occupies the heart of Muslim worship.

Comets, those other bright travelers through the night skies, were likewise greeted with a sense of awe and fear that often spilled over into the realm of the divine. In seventh-century A.D. China, Li Ch'un Feng wrote, "Comets are vile stars. Every time they appear,...something happens to wipe out the old and establish the new." The Crusades were launched in part because the appearance of comets in the skies told the faithful that the time had come to do something about Muslim occupation of the holy places in Palestine. Even in the modern period, comets have been considered portentous and dangerous. Seeing the great comet of 1680—later named after Edmund Halley, who calculated its orbit and predicted it would return in 1758—the theologian Christopher Ness warned that God had sent this cosmic visitor as a heavenly sign that drought and war were soon to follow.

Throughout the nineteenth and twentieth centuries, the scientific Western world dismissed such beliefs as childish nonsense of no more significance than Chicken Little running about in nervous despair, shouting, "The sky is falling! The sky is falling!" The heavens, we knew, were orderly in their movements, with each astronomical body holding to its appointed orbit, much as Aristotle portrayed his universe of crystalline spheres. In the sky, things simply didn't bump into each other.

During the nineteenth century, many scientists began to accept, albeit reluctantly, that meteorites originated somewhere outside Earth. The high nickel content of metallic meteorites differed markedly from the rocks in the areas where they were found and argued for origin somewhere else, beyond our planet. Additionally, a shower of more than 2,000 small meteorites in France in 1803 was investigated by the pioneering astronomer Pierre-Simon Laplace, who declared the rockfall extraterrestrial. Still, such events were seen as little more than curiosities. Until well into the twentieth century, most scientists explained away the obvious craters on the moon as volcanic in origin, not as the result of collisions with objects from space.

Uniformitarianism added to this sense of security. After all, if things are now as they have always been, and if the only meteorites to have survived their collision with Earth are small bits of rock indeed, then there is no reason to believe that astronomical objects pose any danger to our planet. Even as scientists came to accept that meteors had hit the moon and left massive scars, they explained that these events had occurred a long, long time ago, millions if not billions of years in the past. In the safe, settled world of the present, the last threat humans needed to worry about were those errant bits of space-traveling rock, metal, and ice we call asteroids and comets.

The Slow Dawning of Awareness

Looking back, we can be overly harsh about earlier scientists' too-easy dismissal of the danger posed by the skies. Unlike the moon and Mars, which have largely inactive surfaces and show the visible marks of meteors that have plowed into them, Earth's active surface covers over the evidence of impact relatively quickly. As a result, it wasn't until 1957 that scientists even agreed that Earth had a bona fide impact crater. The agreement capped years of sometimes acrimonious debate that ended when geology graduate student Eugene Shoemaker argued convincingly that Arizona's Barringer Crater (now also known as Meteor Crater), which is over 1,100 yards wide and almost 200 yards deep, had been formed during a collision with an asteroid about 125 feet in diameter. Even though the crater was only some 50,000 years old—practically yesterday in geological terms—erosion and other natural processes combined with the absence of a large surviving meteorite, had made identification of the structure as a crater problematical. Shoemaker argued that the crater's pattern of broken rock, iron fragments, and tektites were the footprint of an asteroid impact. Reluctantly science agreed. Space rock had indeed struck Earth hard enough to leave a substantial scar.

With the acceptance of Shoemaker's meteoritic explanation for the Barringer Crater, more impact sites were identified. A classic example is the ring crater over sixty miles across near Manicouagan, Quebec, which has now been flooded by lakes from two hydroelectric dams. Aerial photography showed that the ring might be a crater, and views from orbiting spacecraft confirmed its status as the scar left by an impact approximately 214 million years ago. Yet another very old crater forms the round depression that holds the small city of Nördlingen in Bavaria.

In the past forty years, approximately 150 impact craters, from small to large, have been identified, and the number grows by about three or four almost every year. There is simply no doubt that this number represents only the smallest tip of a very large iceberg. Since crater features weather away, many of the surviving craters are located on the geological features known as shields and cratons, which represent the oldest stable nuclear masses of the continents. The Canadian shield contains twenty-five impact craters, and another nineteen and seven are located in Australia and the Ukraine respectively. Craters on newer continental material have weathered so severely that they are exceedingly difficult to identify. Also, relatively little crater-searching has been carried out in South America, Asia, and Africa; these continents could harbor a great many more impact scars. The floors of the oceans, which cover three-quarters of Earth's surface, are still incompletely mapped and also may contain undiscovered craters. In fact, it wasn't until 1987 that the first ocean-floor crater, a structure more than twenty-five miles wide and estimated to be 50 million years old, was identified in the North Atlantic, approximately 125 miles southeast of the Nova Scotia coastline.

A good example of the difficulty involved in recognizing impact craters after millions of years of geological change comes from Chesapeake Bay. A scientific team headed by Wylie Poag has identified the site where, some 35 million years ago, a large space object hit about 130 miles southeast of where Washington, D.C., now lies. Called a bolide—that is, an object of uncertain composition (perhaps the rock or metal of an asteroid, perhaps the less dense ice, dust, and hydrocarbon tar of a comet) that was somewhere between two-thirds of a mile and seven miles in diameter, traveled faster than a speeding bullet (44,500 to 156,000 miles per hour), and detonated on impact—this object carved a crater twice the size of the state of Rhode Island and nearly as deep as the Grand Canyon. Thirty-five million years later, the crater is impossible to recognize visually because it lies from 950 to 1,600 feet below Chesapeake Bay, its surrounding shoreline, and the continental shelf extending out into the Atlantic. In 1983, Poag and his coworkers aboard the oceanographic ship *Glomar Challenger* collected core samples along the coast of New Jersey that contained the kind of debris, particularly tektites, that indicates a bolide impact. Three years later, cores drilled on land in southeastern Virginia found an immense bed of rock rubble that was actually older than the rocks on which it lay. Apparently the rubble had been dug up and thrown through the air by the bolide's explosion. Fossils in the rubble indicated that it was of the same geological age as the New Jersey tektites. The final and confirming piece of evidence didn't come until 1993, when seismic surveys of deep geological structures by two oil companies seeking out likely drilling sites revealed a large ring crater centered on Cape Charles, a town on Virginia's Eastern Shore.

If an ancient crater of this size located in an area as heavily populated as the mid-Atlantic seaboard takes a good ten years of scientific work to recognize, you can imagine how many more blast scars may lie still undetected and unknown, in both settled and remote regions. We know that, like the moon and Mars, Earth was subject to intense meteor bombardment in our planet's early days. The two oldest known terrestrial craters are Vredefort in South Africa, which is over eighty-five miles wide and 2 billion years old, and Sudbury in Canada, which dates back 1.85 billion years. It is very likely that Earth, which has been around for 4.5 billion years, has even older impact scars, but either they have been completely erased by natural processes or we don't yet have the skill to recognize them for what they are.

Not that the cratering process has stopped. As of this writing, the most recent known impact crater, which measures 165 feet across, was dug out near the city of San Luis in western Honduras on November 22, 1996.

The total amount of material that falls out of space toward

Earth and survives its trip through the atmosphere is staggeringly large. Most of it lands as dust, some as good-sized rocks, and about every 100 million years a massive object strikes our planet. Taking all these sources into account over time, the astronomer Duncan Steel estimates that an average of 200,000 tons of material falls from interplanetary space annually.

This number seems impossibly large, mainly because many meteors come down in isolated regions where the evidence is quickly obliterated. For example, a giant fireball was observed along more than one thousand miles of the Greenland coast on December 9, 1997. The meteor, which burned bright enough to turn night into day more than sixty miles from its path, struck the Greenland ice cap and exploded violently. According to estimates made from seismic data recorded at Cardiff University in Wales, the meteor probably measured between fifty and one hundred yards across, was traveling at about 7,600 miles per hour, and detonated with a force equal to a fifteen-megaton nuclear device. The actual impact site has yet to be found, even though its position was plotted from navigational observations before the impact. The meteor, or what remained of it, probably melted into the ice, which quickly re-formed over it in the deep winter cold and was soon obscured by heavy snow that fell in the hours following the fireball's descent. Similarly, no crater from a fireball and a subsequent explosion in the Australian outback in May 1993 has yet been located.

Reality Check

When the Alvarezes and their colleagues first proposed in 1980 that the impact of a large asteroid led to the extinction of the dinosaurs, many scientists had trouble accepting the notion as realistic. Two facts bothered them. On the one hand, the few truly large craters on Earth, like Vredefort and Sudbury, were very old, dating back to an era when even the most stringent uniformitarian would agree that conditions were quite different from what they are now. As for more modern impacts, none of them was anywhere near large enough to cause even regional catastrophe, much less global extinction. It just didn't seem possible that in our time a space object with planet-destroying dimensions could slam into Earth. It was all too much like Velikovsky, Blavatsky, Churchward, and those other writers with more imagination than science behind them.

Then the universe served up a cold, hard lesson in possibility. Its name was Comet P/Shoemaker-Levy 9.

Named for its discoverers—Eugene Shoemaker, who earlier in his career provided the definitive argument regarding Barringer Crater; his wife, Carolyn; and David Levy—P/Shoemaker-Levy 9* was the ninth comet the three-scientist team discovered. They first found it in March 1993 while working at Palomar Mountain Observatory in California. Subsequent observations from Palomar and other observatories revealed that the comet was in orbit around Jupiter and that it had made a very close approach to the planet about a year earlier. Gravitational forces during this close approach had torn the large comet into a number of fragments, of which six were relatively large (up to about a mile and a quarter in diameter) and a dozen medium-sized, along with a massive array of bits and pieces ranging in size from large boulders to motes of dust. The data also revealed that P/Shoemaker-Levy 9 was on a collision course with Jupiter. For the first time in the history of science, observers had predicted the impact of an astronomical body with a planet and could watch the event happen.

The impact came over a period of seven days in July 1994. Practically every astronomer and major observatory on Earth, as well as the Hubbell Space Telescope, watched intently as the fragments entered the Jovian atmosphere one after another at speeds in excess of 130,000 miles per hour. Heating up into streaking

***The *P* in the name stands for "periodic." Before its capture by Jupiter, the comet had revolved around the sun in an orbit, or period, rather than simply passing through the solar system as an interloper.**

meteors, the large fragments exploded into vast fireballs of immensely hot gas. The gas and debris from each explosion flushed back out the tunnel cleared by the fragment's passage, then expanded, rose, and cooled, forming a large plume. With further cooling, the plume of gas and debris fell back toward the surface of the planet, like the splash of a pebble collapsing into a puddle, heating the atmosphere and producing intense emissions of thermal energy. The biggest show was made by Fragment G, which exploded with a force estimated at 1 million megatons of TNT—about fifty times the nuclear arsenal of Earth. The plume from the fireball rose almost two thousand miles above the upper limit of Jupiter's atmosphere. Dark scars the size of the Earth remained in the usually salmon- and sand-colored atmosphere for weeks afterward, and were visible even in small backyard telescopes.

In its spectacular plummet toward the surface of Jupiter, P/Shoemaker-Levy 9 drove home a number of important points. It demonstrated, for example, that a comet approaching a planet can come apart and strike as a shower of fragments rather than as a single large fireball—a phenomenon that, as we shall see, is very important in understanding the effects of impacting comets on human civilization. It demonstrated another phenomenon scientists had long suspected was true: Comets may explode in the atmosphere before impact, something like a nuclear weapon in air burst. And it showed that, even in contemporary times, a comet could indeed collide with a planet with enough force to cause global extinctions.

With P/Shoemaker-Levy 9, uniformitarianism took another step backward. Earth's neighborhood in the galaxy started looking a lot rougher and tougher.

Tunguska: Impact in Real Time

P/Shoemaker-Levy gave us a look at the effects of a planetary collision with a large comet, but because of the distance from Earth to Jupiter, we don't know many of the details of the impact's effects. In the early years of this century, our own planet had a run-in with a much smaller explosive object, a collision that scientists have now had the time to study firsthand.

This event happened in the early morning hours of June 30, 1908. An immense fireball crossed the dawn-reddening sky over Siberia with a roar that could be heard 600 miles away, then exploded with the estimated force of between 500 and 2,000 Hiroshima bombs in a remote region of forests and swamps. A reindeer herder named Vasiliy Dzhenkoul camping close to the edge of the blast site somehow survived, but his 600 to 700 head of reindeer and his herding dogs were incinerated. Even sixty miles away, in the small trading-post town of Vanavara, the blast made a powerful impression. An eyewitness said, "The sky split apart and a great fire appeared. It became so hot that one couldn't stand it. There was a deafening explosion.... [My friend] was blown across the ground a distance of three *sazhens* [about twenty feet]. As the hot wind passed by, the ground and the huts trembled. Sod was shaken loose from our ceilings and glass was splintered out of the window frames."

The blast, which occurred close to the Tunguska River, colored sunrises and sunsets over Russia, Scandinavia, and much of Western Europe for the next several weeks. Even *The New York Times,* in faraway North America, reported the observation of remarkable lights in the sky on the two nights following the explosion and noted that the display was similar to lights seen after the Krakatoa detonation. In Siberia and East Asia, dust cast twenty-five to over forty miles high in the atmosphere by the blast illuminated the entire visible sky; even at night the heavens glowed like daylight. Mount Wilson Observatory in California and the Smithsonian's Astrophysical Observatory reported a marked decrease in the atmosphere's transparency, apparently as a result of the dust load.

The blast also had fascinating geological effects. It affected seismographs at least as far away as St. Petersburg, some 2,500 miles distant. And the Irkutsk Observatory, about 560 miles away from the blast site, reported storms of disturbances in Earth's magnetic field, a phenomenon that has since been recorded during nuclear tests in the atmosphere.

Still, apart from a few reindeer herders and trappers in central Siberia, no one actually saw what had happened. Ignorant of the actual event, scientists attributed the bright atmospheric lights to solar outbursts causing electrical disturbances in the atmosphere. The seismic data were said to be the mark of an earthquake.

The Tungus people, or Evenks, who are native to the Tunguska region, knew better. Certain that angry gods had delivered a punishment, the tribe's shaman-chief declared the blast site an enchanted region no one was allowed to enter. This prohibition, as well as Tungus stories about a vast area of burned and flattened trees, alerted Leonid Kulik, a Russian scientist who had collected meteorite fragments all over Siberia. He wondered what really had happened at Tunguska. Suspecting that a meteor had hit Earth, Kulik resolved to have a look for himself, no matter what the shaman said.

It took him nineteen years to get to Tunguska. Given the turmoil of the First World War, the October Revolution, and civil war in the early years of the Soviet state, it was 1927 before Kulik launched an expedition. Just getting to Tunguska was formidable. Lying in a region of bogs inhabited by daunting hordes of summer mosquitoes, Tunguska could be reached only by blazing a cross-country trail through some sixty miles of forests, streams, rivers, and swamps.

What Kulik saw when he got there was stunning. Before him lay a vast panorama of charred, felled trees that seemed to have burst into flames all at once on that June morning in 1908. There was no obvious crater, yet the blast had an apparent epicenter a little over thirty-five miles across.

Kulik made four more trips to Tunguska, methodically recording the details of the blast site. He mapped the area of fallen trees, which covered a region about half the size of the state of Rhode Island. Inside the blast's epicenter he discovered a number of neatly shaped oval areas he presumed to be minor craters that had filled in over time. Suspecting that the main meteorite lay hidden in a massive swampy area at the center of the blast site, he dragged the swamp and conducted magnetic probes over both the bogs and the oval craters. Remarkably, he found not so much as a fraction of an ounce of the meteoritic iron that would have unmistakably identified the object. Still, Kulik felt certain an extraterrestrial object was responsible for the blast.

The Second World War stopped further work at Tunguska, and Kulik himself died in a prisoner-of-war camp in 1941. Following the war, an engineer and Red Army officer named Alexander Kazantsev wrote a short story suggesting that Tunguska's odd pattern—a flattened, incinerated forest without a crater—indicated a nuclear blast in the air. Since no nuclear weapons were available on Earth in 1908, the only explanation for the Tunguska site, according to Kazantsev, was the crash of an errant alien space vehicle carrying a nuclear power plant. Kazantsev's story kept recirculating, and in 1958 it was reprinted in a popular book titled *Guest from Space*. Gennady Plekhanov, then director of the Scientific Research Institute in the Siberian city of Tomsk, wondered if there was anything to Kazantsev's theory. Even after fifty years, a nuclear blast would leave elevated radiation levels at the explosion site. Plekhanov decided to take a closer look at Tunguska, and he led expeditions there in 1959 and 1960—the first scientific presence in the area since Kulik's last foray before World War II.

Plekhanov and his colleagues scanned for elevated radiation levels. They didn't find them. They searched for meteorite fragments. Like Kulik, they didn't find them, either. Still, despite the absence of conclusive findings, Plekhanov's meticulous fieldwork met with a warm reception among his fellow scientists when he presented an account of the research at Moscow's prestigious Kurchatov Institute of Atomic Energy.

A tradition began. Every summer Russian scientists braved the swamps and mosquitoes to reach Tunguska and collect more information. Since 1989 and the collapse of the Soviet Union, non-Russians have also been allowed in the area, which had previously been closed to foreigners because of proximity to the high-security cities of Tomsk and Krasnoyarsk. The result, after all these years of work, is a richly detailed look at the most recent large-impact site on Earth.

For example, Wilhelm Fast, a Tomsk State University mathematician working at Tunguska since 1960, has assembled an exquisitely precise map of the 850 square miles of fallen, burned trees. Working from the map, other scientists have calculated that the blast occurred at an altitude of three to five miles with a force of ten to twenty megatons of TNT, while the object was traveling from east to west.

The identity of the object remains a perplexing mystery. The absence of any unusual radiation indicates that it wasn't a space vehicle, even though that theory is still actively promoted by Sakura, a Japanese UFO group headed by Kozo Kowai. Curiously, however, the blast did have biological results. Tree growth accelerated close to the epicenter of the blast, and remains accelerated. Biological mutations have increased, not only at the blast site but also along the path of the object over the Tunguska River region. Changes have affected the Rh blood factor of the local Evenks, genetic traits in certain ant species, and the seeds and needle clusters of at least one type of pine. The cause of these mutations remains unknown.

Since the Tunguska object wasn't a spaceship, it must have been either a comet or an asteroid. Even though comets are often described as dirty iceballs, we're not completely sure what they're composed of. They appear to be made up of approximately equal parts of ice, gravel or clay, and heavy hydrocarbons with the consistency of tar or asphalt. Asteroids are usually carbonaceous (something like cosmic charcoal), rocky, or metallic (primarily iron), depending on where they formed in the asteroid belt that lies between Mars and Jupiter. Asteroids are denser than comets, but they travel at much slower speeds.

To date, no research data indicate definitively whether the Tunguska bolide was an asteroid or a comet. For example, a team of Italian scientists led by Menotti Galli of the University of Bologna has studied chemical evidence in trees that survived the blast. First, the Italian team looked at tree rings for the carbon-14 that would signal the thermonuclear explosion of hydrogen fusing into helium within a comet. They didn't find any. Next, they studied small particles captured within the resin of the trees. The gold, copper, and nickel levels in some of the particles were markedly higher than the levels of the same elements in the surrounding soil. This meant the particles had probably come from the bolide, not from dust kicked up off the ground by the blast. And the metallic particles' smooth texture and round shape indicate that they were heated to extreme temperatures, probably in the heart of the bolide blast. Thus the Italian data point toward an asteroid with a metallic component.

So does a computer simulation developed by three American scientists, Christopher Chyba, Paul Thomas, and Kevin Zahnle. The point of their simulation was to use the laws of physics to develop a description of a bolide that matched the physical data from Tunguska.

Research to date has shown that small meteors, under ten feet in diameter, burn up in the atmosphere, while big ones, over 300 feet across, survive to hit Earth's surface. Midsized objects behave differently, according to the computer simulation. As the bolide rushes through the air, the intense atmospheric pressure on the front end deforms it like pancake batter spreading across a griddle. At the same time, almost no pressure is exerted on the bolide's back end. This difference in forces tears the bolide into pieces, then the pieces, which are subjected to the same force differential, are likewise torn apart. Within seconds, the bolide fragments explosively into a cloud of debris.

The computer showed that a midsized comet, with its lesser density, would detonate fourteen miles or higher over the surface, far too high to fit the Tunguska profile. Similarly, a carbonaceous meteor would explode at nine miles, about four to six miles too high. Unless it reached a speed of twenty-five miles per second, which is rarely if ever seen in nature, an iron-rich meteorite would survive its passage through the atmosphere and strike the Earth, leaving a meteorite—a piece of evidence noticeably absent from Tunguska. But a stony bolide 200 feet wide and falling at a speed of 36,000 miles per hour at a 45-degree angle fit the model. It would explode in the right altitude range, and its blast wave would raise enough dust into the upper atmosphere to cause brilliant sunsets and dawns and light up the nights.

A second model, developed by Jack Hills and Patrick Goda at the Los Alamos National Laboratory, points to the same kind of a meteor. This simulation shows that about 90 percent of the bolide would have burned away, leaving little more than fine gravel to hit the Earth. As a result, no single meteorite of any size remains to be discovered.

A third computer model, however, muddies the waters and shows how difficult it can be to determine exactly what the Tunguska object was. Evans Lyne and Richard Fought of the University of Tennessee and Michael Tauber of Stanford University questioned a key assumption in the simulation done by the Chyba group. The assumption was that the air in front of the bolide reached 45,000 degrees Fahrenheit and that this immense heat was transferred directly to the bolide and burned most of it up. Lyne and his coworkers argued that heat radiated into the surrounding air as well as into the bolide, which reached a high temperature, but much less than 45,000 degrees. As a result, more of the bolide survived to reach closer to Earth before it exploded. In this model, a stony asteroid would explode much closer than three to five miles over the planet's surface, while a carbonaceous meteorite would explode at the right height. So would a comet coming in at an angle much steeper than 45 degrees.

Many of the Russian researchers working on Tunguska hold to a comet as the explanation. They point to the date as important evidence. Earth crosses the Taurid stream, which resulted from the breakup of a comet, twice a year, once from April to June and then again from October to December. The Tunguska event occurred on June 30 during the Taurid passage. And then there's the absence of even small fragments. Years of dragging the swamps have produced not a single piece of rock of indisputably meteoritic origin. And samples of peat taken from the Tunguska bogs and dated to 1908 contain a large number of chemical isotopes that may represent cometary material that fell to Earth when the plume from the blast cooled and collapsed back to the planet's surface. These isotopes may also have originated on Earth, however, so the picture remains less than clear.

In the end, the debate of comet versus asteroid may be a sterile argument. The point is that an extraterrestrial body of uncertain density detonated violently over Tunguska.

Had that blast occurred somewhere other than a remote region with few human inhabitants, the story might have been quite different. Suppose the bolide had streaked through the sky over Paris or New York City or Tokyo. A blast only about one one-thousandth the size of Tunguska killed approximately 140,000 people in Hiroshima, a city that was home to a population of 320,000 to 330,000 people. It seems that a much larger, Tunguska-like blast over a major metropolitan center could kill hundreds of thousands of people, perhaps even millions, and shatter that city's infrastructure. By no means would it be a global catastrophe. To the people in the impact zone, however, that day would seem like the world's last.

Hard Knocks and Splashdowns

Whether an asteroid or a comet, the Tunguska bolide was hardly an oversized space object. Indeed, among the meteors and comets that pass randomly or periodically through Earth's cosmic neighborhood, the Tunguska bolide was at best only a garden-variety object. Thousands of comets and asteroids of much larger dimensions come close enough to Earth to pose a threat of collision. What would happen if one of them hit?

For the sake of example, let's look at the hypothesized effects of an impact with a good-sized rocky asteroid—one a little over six miles in diameter that collides with Earth at a speed exceeding 55,000 miles per hour. Practically the entire mass of the meteor would survive passage through the atmosphere and strike the surface of the planet, where it would release energy in the range of 1 billion megatons—about fifty thousand times the combined nuclear arsenals of Earth.

The atmospheric effects of the impact would affect an entire hemisphere within minutes. Assuming that 10 percent of the energy of the impact went into the blast wave, wind velocity at approximately 1,250 miles from the epicenter would hit close to 1,500 miles per hour, about a dozen times faster than the most powerful hurricane or typhoon, and last for nearly half an hour. Nothing short of a mountain could stand up to such a tempest. Nor could it bear the heat. At the same distance from the blast center, the air temperature would increase by more than 850 degrees Fahrenheit, incinerating buildings, houses, croplands, and forests, melting roadways and bridges, crisping human bodies to crematory ash. Of course, the blast wave would lose energy over time and distance, but even a long way from the explosion's center, the effects would be profound. At a range of approximately 6,200 miles, the wind velocity would exceed sixty miles per hour, last for fourteen hours, and heat the air by over 50 degrees Fahrenheit. Death would be widespread; few could escape.

The blast would also affect the chemical makeup of the atmosphere. In the tremendous heat and pressure of the explosion, atmospheric chemical reactions would lead to the production of poisons, like cyanogen. They would also produce large amounts of nitric oxide, which would soon strip away most or all of the upper-atmosphere ozone that protects the planet from ultraviolet light. Organisms that did survive the blast could soon perish from poisoned air or radiation bombardment.

The other planetary effects of a billion-megaton impact depend on whether the object hits continent or deep ocean. In the case of a continental collision, the force of the strike would set off a seismic wave in the crust that would trigger a worldwide earthquake more fearsome than any living human has experienced. The force would be great enough to affect at least the upper mantle, setting off volcanic eruptions, moving the continental plate, and possibly causing a sudden magnetic reversal—in which case any compasses surviving the holocaust would point south rather than north. If Flavio Barbiero's hypothesis is correct, the planet's spin axis might also move, with the oceans pouring out of some areas and suddenly flooding others, as Strain's calculations suggest.

Most of the energy of the continental impact would be used up in vaporizing the object itself and the land at the site of the collision. The result would be a massive dust plume, estimated at 100 times the mass of the object, totaling hundreds or even thousands of cubic miles and reaching into the upper atmosphere. From there it would spread around the globe over the next two or three months. In about six months, clearing would begin, but the atmosphere would still contain significant amounts of dust for about three years.

Few living things would be around to see the skies return to normal. The heavy burden of dust in the upper atmosphere would act like a thick insulating curtain, keeping out the sun's needed energy. Day would seem like night, and night would be as dark as you can imagine—no stars, no moon. Without sunlight, land plants and marine microorganisms that depend on photosynthesis to survive would die out. The plant-eating organisms and plankton feeders would begin to starve, too, and as they died out, a fatal famine would affect the predators. All the while, the temperature of the planet beneath its veil of stratospheric dust would plummet. Summer would be as cold as winter, and winter colder yet.

An impact point in the deep ocean would have quite different effects. Even in the deepest ocean trenches, the object would carry through the water and strike the ocean floor. The earthquake resulting from the collision would probably be less severe than the one arising from a continental impact, because the crust in the ocean floor is thinner and therefore less strong. A crater would form, on the order of five miles deep and twenty miles across. Meanwhile, the immense volume of water displaced by the impact would rise in a towering column, then fall back, forming a system of tsunamis that decline over distance, then rise up again when they reach the shallow water of the continental shelves. It is likely that waves up to several hundred feet high would strike the continents, sweeping cities before them and flooding vast areas of now-dry land. The entire globe would be affected. The destruction would be less as the distance increased, but it would still be profound.

The impact's effects would not end when the tsunamis finally subsided and the oceans drained back to their new levels. The impact would probably have broken the thin oceanic floor, releasing the hot magma of the mantle, possibly in a violent explosion, perhaps in a slow, sustained venting. Either way, a large amount of sea water would evaporate in the heat and escape into the atmosphere as torrential rain clouds, which would be carried around the globe. A flood, again of global proportions, would be the likely result.

With an oceanic impact, there is no atmospheric dusting to chill the planet. Rather, the release of heat from the mantle and the added evaporation of the ocean turns the climate wet and warm.

Impacts and Ice Ages

The standard explanation of the advance and retreat of continental glaciers by the Milankovitch cycles has a major problem: It doesn't work fast enough. Small, gradualist changes in the orbital parameters of Earth simply can't account for events like the sudden 14 degrees Fahrenheit increase in temperature around 9645 B.C. shown in the Greenland ice cores. Something else must have been involved.

It may have been an oceanic impact, if a suggestive but hypothetical model developed by Emilio Spedicato, a mathematician at Italy's University of Bergamo, proves valid. As Spedicato argues, a continental impact can initiate a glacial period, while an oceanic impact can bring it to an end.

The heavy atmospheric dust veil caused by a continental impact not only shields the sun's light; it also creates a greenhouse effect, blocking the release of Earth's heat into space. Over time, the planet's surface temperature would tend to equalize, even as it declined overall. Since most of the surface heat is found in the oceans, the lower-latitude oceans would cool while the higher-latitude oceans would warm. This heat-exchange process would cause numerous violent storms, which would also arise from rapid atmospheric cooling over the continents in the face of much slower temperature changes over the oceans. Constant heavy rains would fill depressions in the continental interiors—precisely what happened during the last ice age, when Lake Missoula was formed; the Great Salt Lake, Lake Chad in the Sahara, and the Caspian Sea are remnants of these once-extensive continental lakes. At higher latitudes, the storms would deposit snow and ice in a pattern dictated by wind flow. Again, this is what happened in the last ice age. Moisture from the Pacific Ocean was carried eastward across North America, where heavy glaciers formed from the Midwestern area northward. The western shores of the Pacific at the same latitudes, namely eastern Siberia, northern China, and Korea, were relatively free of ice. Similarly, North

Atlantic storms prompted glaciation in almost all of northern and central Europe and western Siberia. Central Siberia endured less glaciation because storms were largely spent when they reached that far inland.

By the time the dust veil fell back to Earth within about three years, a new pattern would have been established: glaciers in the higher latitudes, a colder overall ocean temperature with lower sea level, and reduced evaporation. An ice age would have begun. It would be likely to last, too, owing to the reflection of sunlight off the increased areas of ice and snow and to continued atmospheric dusting from volcanic activity or perhaps even more impacts. An atmospheric phenomenon called diamond powder—droplets of supercooled water in the upper atmosphere—might also result. Diamond powder, currently found only over the poles, reflects sunlight back into space and lowers atmospheric and surface temperatures, adding to the ice-age effect.

An oceanic impact could bring an ice age to an equally sudden end. Tsunamis would wash over much of the continental surface, including the ice-covered regions. Contact with salt water, with its lower freezing point, would initiate glacial melting, in the same way that salt sprinkled on a frozen sidewalk turns ice to water. Warm rains from the water evaporated in the impact and in the subsequent magma flow would speed this melting. Diamond powder in the upper atmosphere would likely melt and fall as rain. With the increased penetration of sunlight and the now-smaller glaciers reflecting less energy back into space, the climate would warm and the glaciers would continue to retreat. The ice age's end would be just a matter of time.

While Spedicato has argued this scenario only hypothetically, two scientists—William Tollmann, of the University of Vienna, and the Russian geologist E. P. Izokh—propose that an impact in circa 10,000 B.C. is the best explanation for the sudden end of the most recent ice age. Among the evidence both scientists cite is the large number of tektites dating to that time period. Tollmann relies, too, on myths of the great flood, like the biblical tale of

Noah, and on a sudden increase in carbon-14 found in fossilized trees dating back to the time period in question. The carbon-14 could have been produced when an impact explosion depleted ozone in the upper atmosphere and exposed the surface of the planet to higher-than-normal radiation. Tollmann is also willing to give the great flood, which he considers one of the results of the ice age's end by impact, a precise date: 9600 B.C., a fascinating (and probably not accidental) coincidence with Plato's time for the sinking of Atlantis.

Rogue Elephants and Savage Swarms: Coherent Catastrophism

According to Spedicato, a one-billion-megaton impact explosion would carve a crater on the order of sixty miles wide. There are at least five structures on Earth with these dimensions: Chesapeake Bay in the United States, Vredefort in South Africa, Chicxulub in Mexico's Yucatán, Popigai in Siberia, and Sudbury in Canada. Clearly, our planet has been struck by space objects of this size more than a few times, very likely with the catastrophic results the Italian mathematician outlines. Yet this focus on outsize objects could be something of an illusion, a misplaced energy that distracts us from the real pattern in our planet's history and the actual threat in our future.

Fixation on outsize objects, however, is slowly becoming the standard paradigm. Until the widespread acceptance of Chicxulub as an explanation of the K-T extinction and the eyewitness experience of watching P/Shoemaker-Levy 9 strike Jupiter, scientists thought Earth immune from astronomical collision. Now that notion is changing, but not all that completely. The new idea is that Earth is struck regularly by relatively small meteorites, which pose little if any threat, and only very rarely by a large comet or asteroid. In this standard paradigm, big objects are like rogue elephants or lone wolves—unusual and unpredictable dangers that come out of nowhere, yet pass away once and for all as soon as they are dealt with.

This standard paradigm underlies 1998's two popular summer

popcorn movies about impacts. In *Armageddon,* Earth is threatened by a massive asteroid knocked out of its usually safe orbit between Mars and Jupiter by the close passage of a comet. In *Deep Impact,* the danger comes from a newly discovered comet whose orbit intersects Earth's. Although the details of the scripts vary, the resolution of the conflict is the same: dispatch into space a brave team that destroys the fast-descending object with strategically placed nuclear weapons. In both cases as well, self-sacrifice of the heroes saves Earth from destruction. With the asteroid or comet put away, Earth can return to its happy ways. The scenario is something like all those old Westerns where, often at the cost of his own life, the good guy destroys the bad guy who has been terrorizing the good townsfolk and a new era of peace and prosperity dawns.

This notion of random, unique threats makes for good Hollywood. After all, no screenwriter wants a threat that keeps on coming—unless, of course, a sequel is in the works. But is the once-in-a-blue-moon, rogue-elephant view of astronomical impact good science?

Objects that can come close enough to Earth to pose a threat—and which are sometimes referred to generically as near-Earth objects (NEOs)—are of two varieties. First are the asteroids. Most asteroids, which are the remnants of a planet that didn't make it to maturity in the early days of the solar system's formation, lie in a belt between Mars and Jupiter. As long as they stay there, asteroids obviously pose no threat to Earth. But the asteroid belt is a fast-moving crowd scene, with asteroids repeatedly banging into one another. When these collisions occur under certain physical conditions and in particular zones of the asteroid belt, material ejected in the collisions can enter new orbits that cross Earth's path. In addition, astronomers are discovering a growing number of asteroids, called Apollo asteroids, that orbit the sun in paths outside the asteroid belt, which could bring them into a collision path with Earth.

Fortunately, the vast majority of asteroids are not all that large. Although a few are hundreds of miles across, most are much smaller, typically less than two miles in diameter, with many no bigger than a pickup truck. Still, such asteroid fragments can lead to dramatic and potentially dangerous incidents, like the meteorite that struck near San Luis, Honduras, in 1996. If it had hit Earth ten hours later, it might have plunged into downtown Manila or Bangkok and killed tens of thousands of people. Luckily, the only loss was the acres of coffee plants that burned in the blast.

The second variety of NEO is comets. Although individual comets have been known and recognized for centuries, only since the 1950s have astronomers determined that a large cloud of comets orbits the sun in the frozen reaches on the far side of the planet Pluto. It is estimated that there are at least as many comets in this distant band, called the Oort cloud, as stars in the Milky Way galaxy. Comets vary markedly in size, with their nuclei estimated to range between two-thirds of a mile and more than 185 miles in diameter. Their orbital period around the sun varies, too. So-called long-period comets take more than 200 years to make one journey around the sun, with some lasting in the neighborhood of 3 to 6 million years. Short-period comets orbit once every 200 years or less.

Until well into the 1970s, astronomers believed that the Oort cloud was stable. It did appear that about every million years a star would pass near enough to churn the orbits of some of the comets, sending most into interstellar space and an occasional one toward the inner solar system. However, these incidents were too weak to disrupt the bulk of the comet cloud. Then, in the late 1970s, astronomers uncovered the existence of cold, dark, very massive nebulae called molecular clouds whose presence makes the Oort cloud unstable and potentially dangerous.

Building on this finding, Victor Clube and Bill Napier, British astronomers who pursue their research at the Armagh Observatory in Northern Ireland, have developed a theory called coherent catastrophism. That is, unlike the rogue-elephant theory of global danger posed by unusual and unpredictable NEOs, coherent catastrophism maintains that predictable astrophysical events regularly send swarms of objects into the inner solar system and place Earth in harm's way.

Two basic mechanisms underlie Clube and Napier's ideas. The first concerns the solar system's passage through the Milky Way galaxy, the small corner of the universe we call home.

When the solar system passes through a molecular cloud, it is subjected to powerful gravitational forces. These affect the sun itself and the planets little, owing to their large mass, but the much smaller comets in the Oort cloud are buffeted and knocked about in a manner that can wholly shift and redirect their orbits. In addition, the sun and its planetary system travel around the Milky Way in a path that periodically takes it into areas affected by tidal forces from the galaxy's spiral arms and the galactic disk itself. These two events—the predictable, periodic passage through stronger tidal fields in the galaxy and the unpredictable encounter with molecular clouds—send comets into new orbits that can direct them out of the Oort cloud. Some of them head far into space, on a collision course with nothing. Others plummet into the inner solar system and possibly close by—perhaps even into—Earth's path.

The second primary mechanism of Clube and Napier's model is the behavior of the comet once it enters the inner solar system. Like P/Shoemaker-Levy 9, large comets tend to break up, with each smaller fragment developing the characteristic head (or coma) and tail of an individual comet. Ancient accounts, including a report by Diodorus Siculus and the records of both Western and Chinese astronomers, report the breakup of comets. The first absolutely reliable recording of such an occurrence, and one whose events illustrate the threats comets can pose, concerned the comet Biela, which was named in 1826. Biela had an orbit of seven years' duration that passed within a mere 20,000 miles of

Earth's orbit; each November 27, Earth came closest to Biela's orbit. On one of Biela's return appearances in late 1845 and early 1846, the comet was observed to split. A faint piece broke off the main comet and rapidly grew brighter. When the comet returned in 1852, the two pieces were 1.5 million miles apart. Then, in 1865, when Biela and its companion should have reappeared, they were nowhere to be seen. On November 27 of that year, Earth ran into a meteor swarm that lit the sky, according to one observer, "like a cascade of fireworks." An estimated 160,000 shooting stars streaked through the sky during the six hours the display lasted. Called the Andromedid shower, the meteors returned for another spectacular show on November 27, 1885. The Andromedid shower still exists, although now it is largely spent and the fragments have been distributed over the entire orbit of the former comet. Neither Biela nor its companion was ever seen again.

Comet fragmentation apparently results primarily from internal forces within the comet's coma. Gravitational pull from the sun can fragment a comet, much as Jupiter fragmented P/Shoemaker-Levy 9, but since breakup has been observed at many points along cometary orbits, not just close to the sun, other pressures must also be at work. Perhaps thermal forces crack the comet's gravel and ice and fracture it. Or perhaps it runs into other space debris and breaks into pieces in the collision.

It also appears that as comets orbit through the inner solar system they do not simply evaporate away into gas—which, of course, poses no danger to Earth. When Halley's Comet made its most recent passage through the inner solar system, it was intercepted by a number of spacecraft that gave us our first close-up look at a comet. The escaping gas that formed the tail came from only a few jets on the surface, which had the solid black look of a large cinder. As even more of the gas escapes, the comet will eventually take on the appearance of a soot-colored asteroid. In fact, at least some of the Apollo asteroids, which orbit in the solar system outside the asteroid belt, may be the spent remnants of comet cores.

Clube and Napier hypothesize that when a comet from the Oort cloud enters the inner solar system and assumes a new short-period orbit, it fragments again and again, forming a long, wide stream of debris. After perhaps 10,000 to 20,000 years, the comet has turned into an asteroid of substantial size. This fragile inner core is gradually sliced, carved, and whittled into various bits and pieces, both through internal forces and collisions with smaller objects. Throughout its life span in the inner solar system, the comet creates a large, ever-growing, ever-widening stream of fragments.

The results of a planetary body's passage through such a stream were observed on the moon in late June 1975, thanks to the aid of seismometers on the satellite's surface. As the moon encountered the Taurid meteor stream, its surface was struck with as many one-ton space rocks in five days as it had been in the prior five years. Nothing equivalent was seen on Earth because the meteors came in on the daytime side of the planet and burned up invisibly in the atmosphere. Had the stream encountered Earth at night, the show of shooting stars would have given us some idea of the Andromedid spectacle of 1865.

Because short-period comets keep fragmenting and fragmenting as they orbit the sun, they eventually turn into a band of dust that spreads out over the entire orbital path. Most likely, comets pose the greatest danger early on in this fragmentation process, when the stream is relatively narrow and the comet pieces larger rather than smaller. Should Earth pass through such a young stream, it would be struck with showers of fragments, some of them big enough to do serious damage. Even fragments of a size too small to cause more than local disaster, striking in clusters of impacts across the face of the globe, could lead to serious cumulative destruction. And the comet's dust infused into the upper atmosphere and added to the debris injected into those same regions by impact would add to the planetary cooling effect of continental collisions.

Meteor streams resulting from a comet's slow fragmentation have yet another ominous characteristic: Year after year, Earth passes through them. Some years, of course, the planet may dodge the bullet, or more accurately the meteor. But in other years, swarms of space rock, ice, and tar could come speeding through the atmosphere toward the surface of the planet.

Clube and Napier note that the Taurid meteor stream, Comet Encke, and several large asteroids all follow orbits that appear related. In addition, this orbital pattern follows a broad tube of dust and debris that Earth enters in April and from which it emerges in late June, and then reenters in October and escapes again in December. As Clube and Napier see it, all these objects, from Encke to the smallest Taurid meteors, came from the fragmentation of a very large comet over a period of perhaps 20,000 years. Backtracking from current data gives us some idea of what the original comet was like and what path it may have followed. Given the mass of comet, asteroids, meteor streams, and debris, the object must have been more than sixty miles across in its early days. Also, Comet Encke and the large asteroid called Olijato had nearly identical orbits some 9,500 years ago. Very possibly the comet, or one of its large fragments, underwent a massive disintegration at that time, with Encke and Olijato representing two of the largest surviving remnants of that particular fragmentation. There may have been others. From 3500 to 3000 B.C., Olijato had repeated close encounters with Earth. Any sibling objects traveling on the same path might well have struck the planet during the same period. And the Tunguska object, with its date of June 30, right at the time when Earth was exiting the edge of the Taurid stream, may well have been the most recent of these large fragments to hit the planet.

According to Clube and Napier, it takes comets some 3 to 5 million years to tumble from the Oort cloud into the solar system.

During this period of time a giant comet trapped in an Earth-crossing orbit creates a large swarm of asteroids. The giant comets recur in Earth-crossing orbits at intervals of about 100,000 years, and our planet's encounters with the asteroid stream of the comet peaks roughly every millennium. Depending on the stream's orbit, the danger to Earth would be greatest at one or two times of the year for a few hundred years. Some years nothing would happen; in others the bombardment over hours or days would be intense. Intermittent periods of calm between the bombardments would last perhaps 1,000 years, when the cycle could begin again.

Clube and Napier suggest that we are currently in the tail end of such a 3-to-5-million-year episode, when comets have been passing through the inner solar system. But current conditions—Earth's positioning near the galactic plane of the Milky Way, our closeness to one of the galaxy's spiral arms, and recent passage through molecular clouds—could well initiate a new flow of comets from the Oort cloud into orbits that bring them close to Earth. The future, Clube and Napier predict, does not bode well.

Craters and Smoking Guns

One of the primary scientific benefits of Clube and Napier's model is that it provides a mechanism for explaining the how, when, why, and what of impacts. When the Alvarezes and their colleagues first proposed a collision with a comet or asteroid as the cause of the extinction of the dinosaurs, the seeming randomness of such a collision made the idea suspect in the eyes of many scientists. In a manner of speaking, they were hypothesizing the astronomical equivalent of a piano falling on your head while you're walking along Fifth Avenue. There you are, just minding your own business, when this half-ton mass of mahogany and ivory plummets out of the sky and squashes you into the pavement. Scary, for sure, but without knowing why the piano happened to show up in the same place you did, basically unpredictable. Scientists have trouble with this kind of scenario. We prefer models that deliver the ability to predict and analyze. In science, even chaos has its theory.

Clube and Napier added something to the impact model for the K-T extinction by providing the missing theoretical background. Their hypothesis of cometary flux into the solar system as the result of galactic tides and run-ins with molecular clouds is plausible.

Of course, science requires evidence as well as elegance for hypotheses. Which raises the issue: Are there any data to support Clube and Napier's notion of clustered impacts resulting from comet fragmentation as the cause of disaster?

The K-T extinction is just one of five major extinctions, and it was by no means the biggest. That one happened at the boundary between the Permian and the Triassic periods, about 245 million years ago. In relatively few years, life in the seas came almost entirely to an end. Something on the order of 90 to 96 percent of all marine species were wiped out. The ax fell on creatures at every level of evolutionary complexity: all the trilobites, 500 species of plankton, most of the sponges and echinoderms (starfishes, sea urchins, and their relatives), and hundreds upon hundreds of species of corals. Ten million years passed before coral reefs, those colorful and highly productive submarine structures of tropical seas, made their reappearance. The extinction of land-based animals was equivalently catastrophic. Approximately 80 percent of all the amphibian and reptile families then extant disappeared.

Like the K-T boundary, the Permian-Triassic transition layer contains far more than the normal terrestrial amount of iridium. To date, no large crater on the order of Chicxulub has been found that corresponds to the Permian extinction. Could it be that a swarm of smaller fragments hit over a period of hours, days, weeks—even years or decades—clogged the seas with debris, and filled the atmosphere with a chilling veil of comet dust?

Interestingly, two chains of craters contemporaneous with other major extinctions have recently been identified. One, announced by the National Aeronautics and Space Administration in 1996 and located in the central African country of Chad, was revealed by radar images taken during the missions of the space shuttle *Endeavor* in 1994. Focused on a known impact site named Aorounga in northern Chad, the images showed two other nearby structures that also appear to be craters. Dating to 360 million years ago, the craters may well have resulted from the fragmentation of an asteroid or comet that struck Earth in pieces. At the time of the impacts, Earth underwent a mass extinction during what is known as the Devonian-Carboniferous boundary. Like the K-T boundary, the one marking the Devonian-Carboniferous transition contains an unusually excessive amount of iridium.

"The impacts in Chad weren't big enough to cause the extinction," said Adriana Ocampo, a Jet Propulsion Laboratory geologist, at the time of the announcement, "but they may have contributed to it. Could these impacts be part of a larger event? Were they, perhaps, part of comet showers that could have added to the extinction?"

The second chain, identified only in 1998, dates to approximately 214 million years ago, around the time of the Triassic-Jurassic boundary, an extinction every bit as catastrophic as the K-T event. The research, conducted by David Rowley of the University of Chicago, John Spray of the University of New Brunswick (Canada), and Simon Kelley of The Open University (United Kingdom), showed an unsuspected connection between five craters spread across two continents. At the time when the craters were formed, North America and Europe were in much different positions than they are now. When the researchers moved the continents back to their ancient geographical locations, the relationship among the craters was striking. Three of the five craters—Rochechouart in France, and Manicouagan and St. Martin in Canada—lay on the same latitude of 22.8 degrees, forming a chain now some 3,100 miles long. The other two craters lined up on identical declination paths with two of the craters in the first chain: Obolon in Ukraine with Rochechouart, and Red Wing in Minnesota with St. Martin. The chance that the craters could be randomly aligned this way is near zero, according to the three scientists' analysis.

All in all, the craters' arrangement suggests a comet that broke into at least five fragments and hit Earth in two groups of two, followed by a single piece: Rochechouart and Obolon, St. Martin and Red Wing, then Manicouagan, at more than sixty miles in diameter the largest of the bunch, coming down all alone. Possibly there were more than five fragments. At the time of the impact, Earth at those latitudes was mostly ocean, so other pieces may have fallen into the sea.

At least one other iridium layer is known, positioned at the boundary between the Ordovician and the Silurian layers. As yet, no corresponding crater has been found. But since the boundary lies 440 million years ago, erosion and other geological processes would make recognition of an impact structure this old very difficult.

Nemesis: The Death Star Alternative

Clube and Napier posit that the two forces driving the influx of comets into the solar system are contact with molecular clouds and galactic tides. Another model for explaining how it is that comets come hurtling into Earth's way has been offered by Richard Muller, a physicist at the Lawrence-Berkeley Laboratory on the University of California campus.

Fascinated by dinosaurs since childhood, Muller was intrigued by the impact theory being developed by his colleagues at Berkeley in the late 1970s and 1980s, but he was involved in other research projects at the time, particularly in high-energy physics. Then he came across a scientific paper by David Raup and J. John

Sepkoski Jr., of the University of Chicago, which argued that the mass extinctions have occurred approximately every 26 million years in the fossil record. This notion intrigued Muller because it meant that the K-T extinction the Alvarez team was studying was not random and unique, but part of a periodic pattern. A pattern, though, meant a mechanism—and what could that be? Which predictable, repeated astronomical event could send a stream of comets into the solar system every 26 million years or so?

After trying and discarding hypothesis after hypothesis, Muller had what he describes as a "eureka" moment during a conversation with orbit dynamics expert Piet Hut. The sun, Muller realized, could have a twin, most likely the type of star known as a red dwarf. The twin orbits the sun in a path that jitters slightly because of the gravitational influence of other passing stars. About every 26 to 30 million years, the twin star comes close enough to the Oort cloud to send a shower of comets into the solar system and set off another mass extinction. Muller named the twin star Nemesis, after the Greek goddess who brought down the rich, powerful, and proud.

Since then, Muller has been conducting a painstaking star-by-star search for Nemesis. According to Muller's model, Nemesis follows an elliptical orbit that brings it a little less than one light-year away from Earth at its closest point, almost three light-years away at its farthest. While virtually all the red dwarf stars in our galactic neighborhood have been identified—there are about 3,000 of them—few have had their distances measured. Muller is looking for the one that lies where he predicts it must, between approximately one and three light-years away.

Muller's hypothesis is interesting, and he himself is a fascinating man, full of energy and enthusiasm for the wild frontier of astrophysics he is exploring. His model, though, is open to question on two grounds.

One is the 26-to-30-million-year periodicity that Raup and Sepkoski have proposed. Many other paleontologists disagree with their conclusion, which is based on complicated statistical analysis of the sort only specialists can fully appreciate. Extinctions have been repeated, the other scholars argue, but not with the specific periodicity Raup and Sepkoski maintain. A pattern more variable than the 26-to-30-million-year periodicity argues against the existence of a twin star.

The other issue, one addressed by Clube and Napier, has to do with the supposed distance between the sun and Nemesis. Twin stars, or binaries, as they are also known, are quite common in the galaxy. Among binary stars of the same age as the sun, the distance between the twins averages around 5,000 astronomical units (an astronomical unit is the distance from sun to Earth, or approximately 93 million miles). According to Muller's model, Nemesis lies an average distance of 90,000 astronomical units from the sun, and at its farthest orbital point is 180,000 astronomical units away. In other words, Muller is proposing a system that has not yet been observed within the known universe.

None of this analysis, like the still incompletely understood geological record of impacts on Earth, resolves the debate once and for all. Still, when I look at what we now know, I find Clube and Napier's case persuasive. And when their model is used to explain a variety of seemingly inexplicable events in human history, it becomes even more convincing.

A Twisted Moon and Chaos in the Pacific

Clube and Napier themselves rely on another body of evidence, one we have turned to again and again: mythology and folklore. Practically every mythological tradition recounts stories of terrible cataclysm, often marking the shift from one world-age to another. Zeus battles Typhon; the gods of Olympus take on the Titans; floods wash away all save Noah, Deucalion, and Utnapishtim; fire and brimstone descend upon Sodom and Gomorrah; God casts Satan from heaven, launching the "dubious battle" John Milton depicts in *Paradise Lost;* Valkyries and Furies fly over the surface of the Earth and threaten horrible death to evildoers; the hero Marduk takes on the fearsome ocean monster Tiamat and creates a new universe from her torn body. Again and again, order battles chaos, with the fate of humankind and the world itself often at stake.

As Santillana and Dechend show in *Hamlet's Mill,* myth reveals considerable evidence that ancient stories were used to mark precession and the passage from one world age to another. Yet precession is a slow phenomenon, not one that of itself would inspire the fear obvious in the many mythological battles of order and chaos against a background of destruction and cataclysm. Were ancient people so anxiety-ridden and nightmare-beset that the glacial movement of the stars gave them the shakes? Or did their memory include celestial catastrophes as terrible as the ones described in their stories, disasters they later associated with the slow, inevitable movement of the heavens?

It happened that on the night of June 25 in A.D. 1178, several monks of Canterbury in England were watching the horned moon as it became visible after sunset. Suddenly the upper horn of the moon split in two, with a torch of flame rising at the point of the split. The body of the moon wavered and writhed "like a wounded snake," in the words of the monk Gervase, who chronicled the event. A dozen times the same phenomenon of flame and writhing occurred, until the entire moon took on a black appearance.

In medieval times, visions were considered as real as the ordinary events of everyday reality. Against such a cultural backdrop it is easy to dismiss Gervase's account as the further imaginings of the overly religious.

The meteorite expert J. B. Hartung wasn't so sure. He argued that the monks had seen some kind of astronomical body collide with the moon. The torch of flame shooting up was either a cloud of incandescent gas from the blast, or sunlight reflected in the dust rising from the crater. The blackening of the moon's surface was also the result of dust, which was carried over the lunar surface in the short-lived atmosphere created by gases released in the explosion. Any crater, Hartung argued, would be at least seven miles across, have bright rays of ejected material extending out for seventy miles, and be located within a specific area right at the edge of the moon or just over into the dark side.

One crater, discovered well after Hartung's prediction by one of the early lunar missions photographing the moon's back side, fits the bill. Named Giordano Bruno, for a sixteenth-century heretic burned at the stake for his theological views on astronomy and his impolitic manner in broadcasting them, the crater is thirteen miles in diameter and has a system of brilliant rays reaching out for several hundred miles. It lies in the lunar region where Hartung said it must, and the brightness of its rays indicates that Giordano Bruno is a very new structure.

Further evidence about Giordano Bruno's recent origin comes from the work of the lunar astronomers O. Callame and J. D. Mulholland. The moon does spin on its axis, but the period of spin matches the period of its orbit around Earth, so that we always look at the same lunar face. Callame and Mulholland calculated that a collision with an object large enough to excavate Giordano Bruno—on the order of 100,000 megatons—would cause the moon to wobble several yards on its axis over a period of about three years. Laser-ranging data collected at the McDonald Observatory at the University of Texas since the early 1970s show that the moon does indeed wobble about fifty feet on its axis every three years. Since this kind of motion slowly dies out over about 20,000 years, the wobble can be explained only by a recent major impact.

The date of the 1178 lunar impact—June 25—is close to the Tunguska date of June 30. There's more behind this close identity than mere coincidence. The end of June comes at the time when Earth is crossing out of the Taurid meteor stream, and the year 1178 came during a century when the Taurid stream rose to a peak.

During that century, large comets made repeated frightening appearances in the skies over Europe. When one of these huge comets appeared during a meeting of bishops, the prelates decided that God was threatening humankind with destruction for its sins, and resolved to mount the First Crusade in order to win divine favor. Similarly, Chinese astronomers in the same period noted a sharp increase in comets and fireballs around A.D. 1150, when the number of these celestial objects leapt to ten times the normal background number.

If a major object hit the moon during the late twelfth century, and if this period was marked by an obvious increase in the number of comets flying by Earth, it would seem likely that some of these space objects collided with our planet. Clearly there was no extinction-level event like Chicxulub. Still, a cluster of impacts on the order of the Tunguska bolide could have had powerful consequences for human civilization.

Emilio Spedicato argues that the historical record for the Pacific Basin in the late twelfth century shows a pattern of inexplicable chaos. For example, the Maoris of New Zealand disagree with the Western idea that the island's moa birds disappeared because of overhunting. According to their legends, fire fell from the sky several centuries ago and burned up the forests where the birds lived. There is physical evidence to support the Maoris' contention. A series of recent shallow impact structures called the Tapanui Craters have been found on the South Island and dated to about 800 years ago, or circa A.D. 1200. Geological samples show layers of soot dated to the same period as the craters and provide evidence of extensive fires of the sort the Maori stories describe.

During the same time, Polynesia was swept by waves of migration. Apparently whole peoples pulled up stakes and moved on to new locations, in some cases conquering and then absorbing the groups who already lived there. Royal lines were disrupted, new gods replaced old, customs changed, and novel religious practices arose. All these events happened suddenly and without apparent explanation in the twelfth century.

South America also underwent rapid change. Mochicas, Chimus, and the other great cities along that continent's Pacific coast had built a remarkable civilization. Among other accomplishments, these people created a complex system of irrigation canals and erected huge pyramids of compacted soil. The largest of the pyramids, called Tucume and located near the modern Peruvian town of Lambayeque, was only a little over 200 feet high, but it contained one-third more volume than the Pyramid of Khufu (Cheops) at Giza. Then, during the twelfth century, the coastal civilizations suddenly collapsed, the cities turned into ghost towns, and the survivors did not return. Rather, a new civilization, the one we know as the Inca empire, arose in the high Andes and built a remarkable system of roads that ran from Colombia to Chile. It could be that the coastal civilizations collapsed because of tsunamis resulting from comet impacts, and that the survivors took themselves and their construction skills to the high mountains, where they sought refuge well above the level of even the angriest sea.

In an uphill movement similar to the one in South America, the people who would later be known as the Aztecs also migrated, again without apparent explanation. They moved from a location on the Pacific coast of Mexico known as Aztlán, probably near the modern city of Mazatlán, to the high-altitude, mountain-enclosed valley where Mexico City now lies. According to recent historical studies, this migration occurred in the middle of the twelfth century. Like the Incas, the Aztecs may have moved in order to avoid tsunamis. And their custom of human sacrifice could have been a way of warding off the ill will of the gods, which they saw as responsible for the terrible catastrophe that had earlier befallen them.

In northern China, the second half of the twelfth century witnessed the beginning of the end for the corrupt Song dynasty under pressure from the invading Juchen tribes from Manchuria. Weather conditions at the time were unusually severe. In A.D. 1194 the Yellow River flooded so catastrophically that it almost completely destroyed the northern Song capital of Kaifeng. The river itself shifted its outlet to the sea, moving from a location to the north of the Shandong Peninsula to one hundreds of miles south.

Deeper inside Asia, all was turmoil. According to an account by Khubilai Khan, his grandfather Genghis Khan, as an adolescent, saw a sign in the sky—very possibly a comet—in the late 1170s or early 1180s and took it as a portent signaling his own rise to power. Led by Genghis Khan, Mongol horsemen swept out of their high, dry plateau and conquered all of Central Asia, Persia, Afghanistan, and a portion of India, carving a path of death and destruction rarely equaled in human history. Typically the Mongol invasion is explained as the result of the extraordinary ambition of Genghis Khan and his desire to avenge a long list of wrongs done to himself, his family, and his people. Yet some evidence suggests that the Mongols, who inhabited a land that is harsh at the best of times, had been pushed to the brink by a serious deterioration of the climate. The Persian chronicler Al Juvaini noted that in A.D. 1260, apples could again be grown in Mongolia, a feat that had been impossible for the prior two generations. This report points to winters in Mongolia so severe that ordinary agriculture had become impossible, forcing the Mongols to expand outward or starve. The folk tales of the Yakut people, who live farther north in eastern Siberia, tell of an evil star with a tail that means storm and frost even in summer. This tale could well be the record of a cosmic winter effect caused by the impact of a number of bolides in the region, perhaps even the same series of events that put the Mongols on their horses and sent them West in search of warmer, more fertile lands.

Mongolia is also thought to have been the site where the Black Death, which killed about one-third of the population of Europe in the middle of the fourteenth century A.D., arose. The causative agent of the plague, *Pasteurella pestis,* followed the caravan routes out of Asia and made its way with traders into the ports of the Mediterranean. In an indirect way, the Black Death may have owed its origin to the ecological disruption affecting Mongolia a century and a half earlier.

None of this evidence is definitive. Still, it points to the possibility that meteor swarms could have come down in and around the Pacific during the twelfth century, causing tsunamis and upsetting weather patterns so completely that whole civilizations collapsed or moved.

To the Bronze Age and Before

The widespread, seemingly inexplicable chaos marking the late twelfth century A.D. in the Pacific is hardly a unique event. The end of the Bronze Age in circa 1200 B.C. is another. And approximately one millennium earlier, an equivalently widespread collapse of civilizations affected the Middle East, South Asia, and the Far East. In a period spanning five centuries from 2500 to 2000 B.C., several major civilizations met a sudden, unexplained demise: Early Bronze Age societies in Israel, Anatolia, and Greece; the Old Kingdom in Egypt; the Akkadian empire in Mesopotamia; the Himland civilization in Afghanistan; and the Indus Valley civilization in Pakistan. Much like the collapse marking the end of the Bronze Age, a dark age followed. Take Egypt as an example. The few pyramids built during the Middle Kingdom, which rose from the remains of the Old Kingdom, are but pale shadows of the monuments at Giza. Their engineering is sloppy, the workmanship poor, the art hackneyed. Egyptian culture had suffered a mighty setback, almost as if it had been the victim of a frontal lobotomy. Whatever events caused Egypt such damage, they were mighty indeed.

Clube and Napier maintain that during a period when a large comet is breaking up in an Earth-crossing orbit, meteor swarms are likely to peak at approximately millennial intervals. If we start with the twelfth century A.D. and the serpent writhing on the moon and go back a thousand years, we find ourselves in the second century A.D. The Roman Empire was beginning to crack and crumble, and religious enthusiasm was sweeping the Mediterranean as waves of true believers, some of them Christians, prophesied the fast-approaching end of the world. Events become even more dramatic a little over a millennium earlier, close to the sudden collapse of the Bronze Age throughout the eastern Mediterranean. Leap back yet one more millennium and a bit more, and civilization is again collapsing from the eastern Mediterranean coast all the way to China. Clearly, there is something to the roughly millennial interval Clube and Napier propose.

As yet, no large crater dating to 1200 B.C. or thereabouts has been discovered in the eastern Mediterranean. The Bronze Age, though, could have collapsed not from one big knockout blow, but from a series of celestial left jabs to the jaw. Suppose that for a period of several decades, many Junes and many Novembers were marked by fireballs of the Taurid stream streaking out of the skies toward Earth. Some of them would have detonated in the atmosphere, creating shock waves of intense energy and heat, the kind that could have swept away a stone-walled city and burned its rubble to cinders. Others would have landed in the sea, raising fearsome tsunamis that could obliterate coastal settlements and cities. And yet others would have struck the surface. It wouldn't take a direct hit to destroy a city. Hot material ejected from the blast site could set fire miles away to cities, villages, and croplands, much as the myth of Phaethon describes. That kind of hard blow to the crust, in a region as seismically and volcanically active as the eastern Mediterranean, might trigger earthquakes and initiate lava flows. And then there would be the dust, rising in a plume into the upper atmosphere, cooling the region for days or weeks like a veil drawn across the sun.

Some evidence from the end of the Bronze Age fits this model. People abandoned settlement sites that had been in use for centuries. The attacking hordes that beset Egypt and spelled the eventual end of the New Kingdom appear to have been part of a general movement of people from northern latitudes to more southerly locations, possibly a response to unusually severe winters. And a steep drop in the growth of bog oaks in Ireland during the years 1159 to 1140 B.C. points to a dramatic deterioration of climate of the sort that would accompany a cosmic winter.

Similar evidence is mounting around the earlier Bronze Age collapse in the late third millennium B.C. Researchers have found unambiguous evidence of major climatic upset right around 2300 B.C., give or take a couple of centuries. Areas like the edge of the Sahara and the plain around the Dead Sea, which had been farmed, dried into deserts. Tree rings show disastrous growth conditions in circa 2350 B.C., and sediment cores from lakes in Europe and Africa indicate a massive drop in water level. Large regions in Mesopotamia were devastated, flooded, or burned. Something very big was going on. And, I think, it came from the heavens, in the form of fragments from a comet that had been sent our way from the Oort cloud millions of years before.

Centuries ago, the English poet John Donne wrote that no man is an island. Neither, it seems, is Earth.

7

Learning from the Past, Looking Toward the Future

OF ALL THE CATASTROPHES THAT CAN DESTROY CIVILIZATION, the most dangerous and destructive are asteroid and comet impacts. Space objects have the capacity to cause widespread devastation and loss of life on their own, and they can trigger follow-up events, such as earthquakes, vulcanism, drastic climate change, accelerated tectonic plate movements, even sudden shifts of the poles, that add to the overall level of destruction. Obviously this danger isn't simply theoretical. The two Bronze Age catastrophes and the chaos in the twelfth-century A.D. Pacific point up what can happen when the Taurid stream sends its fragment swarms in Earth's way. It may even be that the advanced civilization that carved the original Great Sphinx of Giza and then vanished was wiped out in this very same manner.

The understanding that asteroid and comet impacts have profoundly affected the history of life and civilization has two important ramifications. One is philosophical. As I pointed out in the beginning of this book, the paradigm from which science views the world is changing rapidly and completely. In the not-so-distant past, biologists looked to the laboratory for their answers, geologists looked to Earth, astronomers looked to the sky, and historians looked to the human events of politics and culture, as if biological, geological, and certainly astronomical reality were mere stage props that had nothing to do with them. No longer does such a divided worldview work. If civilization can rise or fall with events in the Oort cloud 1.5 light-years away, then certainly our world is larger, more perilous, and more interconnected than we had ever imagined.

The second ramification is practical. If we extrapolate the thousand-year intervals Clube and Napier propose, it would appear that, except for the occasional wayward bolide like the Tunguska object, Earth is currently enjoying a quiescent period. But around 2200 A.D. it is likely that a new flow of comet fragments will enter Earth-crossing orbits and pose a real threat to our planet. This prospect grows even more ominous when we consider that changes currently affecting life on Earth, namely stratospheric ozone depletion and global warming, would augment the effects of bolide impacts. As bad as the catastrophes were that befell Earth in the past, future disasters could be much worse.

Living and Nonliving Together

An important part of the new developing paradigm, and the framework from which we can best understand the implications of major asteroid impacts in our world, is the Gaia hypothesis. Developed largely by the pioneering work of the atmospheric scientist James Lovelock and the microbiologist Lynn Margulis, the Gaia hypothesis proposes that the physical and chemical conditions of the Earth—including the planet's surface, the atmosphere, and the oceans—are formed by living organisms to create conditions favorable for life. Where Darwin and his fellow evolutionary theorists look at life as an adaptation to the physical and chemical demands of the environment, Lovelock and Margulis maintain that life itself has shaped the environment. The implications of the Gaia hypothesis are enormous.

In the history of Western science, the foundation of the Gaia hypothesis stretches back at least as far as James Hutton (1727–1797). As we saw in chapter 1, Hutton envisioned Earth as a great recycling machine whose purpose was creating an environment life could inhabit. Rock weathers into soil, which is then eroded away and carried by water into the oceans, where it accumulates into sediments, which eventually compress into rock that is raised above sea level to begin the cycle again. Since soil and marine sediments are critical to all living things, life depends on the continuous recycling of nonliving rock.

Lovelock took this idea much further, although he had no such intention at the time. During the 1960s, Lovelock served as an instrumentation consultant to NASA's Jet Propulsion Laboratory in Pasadena, California. One of JPL's projects was assessing the likelihood of life on Mars. Most of the technological answers being explored by JPL scientists entailed taking a series of samples from the Martian surface and analyzing them for either microorganisms or chemicals whose presence would signal the prior or current presence of life. Lovelock wondered about this approach. How, he asked, could we be sure that life on Mars followed the Earth model? The more Lovelock considered that question, the more he realized he was actually asking something deeper: What, after all, is life?

Increasingly, Lovelock saw that the answer had to do with entropy, one of the most confusing and misunderstood concepts in physics. According to the Second Law of Thermodynamics, all energy will eventually dissipate into heat and be unavailable for useful purposes to either machine or living organism and thus increase entropy, which stands for randomness or chaos. Life works against entropy, by organizing energy in ways that counter the constant running-down into chaos. When an animal or plant dies and its body decays, entropy increases. But while the organism remains alive, its activities counteract entropy. Find a site where entropy is other than what the laws of physics and chemistry would predict, Lovelock reasoned, and you are looking at the evidence of life.

Following this line of thought, Lovelock studied what it was that made Earth different from Mars and Venus. An atmospheric scientist by profession, he paid particular attention to the atmospheres of the three planets. Earth's air, which is a mixture of gases, has a composition markedly different from the blend constituting the atmospheres of the other two planets. Our atmosphere is approximately 20.9 percent oxygen and a little more than 0.035 percent carbon dioxide. On Mars and Venus, carbon dioxide exceeds 95 percent and oxygen is barely present. According to the laws of chemistry, Lovelock realized, Earth's atmosphere shouldn't be what it is. If the nonliving components of Earth's surface and its atmospheric gases were placed in an immense bottle and subjected to the same level of sunlight that falls on our planet, they would react chemically with one another, always in a way that increased entropy. Over time, the result would be an atmosphere very much like that found on Mars and Venus—mostly carbon dioxide, with only a trace of molecular oxygen.

Earth's atmosphere maintains its curious and unpredictable combination only because of the presence of life, Lovelock deduced. We have our air because plants and photosynthetic microorganisms take in carbon dioxide and give off oxygen. The carbon dioxide–oxygen cycle is only one of many biological mechanisms that work something like a thermostat to regulate atmospheric gases within the zone life needs in order to exist.

An example of this process, technically called homeostasis, arises after a major impact, like that of Chicxulub. The huge forest, brush, and grassland fires resulting from the blast released massive amounts of carbon dioxide into the atmosphere. At the same time, however, the immense quantities of upper-atmospheric dust resulting from the explosion killed off a substantial number of photosynthesizing organisms, both land plants and marine organisms. As a result, less of the now more abundant carbon dioxide was being used, and less oxygen was being produced. The atmospheric concentration of the one gas rose, while that of the other fell. Surely some plant species that had survived the immediate effects of the impact became extinct in this atmosphere, while others somehow survived and perhaps even took advantage of opportunities presented by the changes. As these survivors evolved in the brave new world of the post-impact era, their population rose, more carbon dioxide was consumed, and more oxygen was produced. Over time—likely millions if not tens of millions of years—the atmospheric system followed a life-driven route back toward its original steady state.

This is homeostasis, the sum of many complex interactions between the living and the nonliving that contributes to conditions suitable to life. Homeostasis guides the system back toward a desirable range even in the face of an upset as large as an incoming comet.

When Lovelock was developing his original formulation of this idea, he didn't really know what to call it. He toyed with various complicated monikers drawn from scientific jargon, but none of them seemed to fit. Then one of his neighbors with a pronounced gift for words, the Nobel Prize–winning English novelist William Golding, author of *Lord of the Flies,* proposed Gaia, the name of the ancient Greek goddess who symbolized Earth itself. Lovelock liked the association, and the name stuck.

Naming the hypothesis after a goddess has given it a certain cachet among New Age types who tend to misunderstand the concept in their rush to use it for their own purposes. The Gaia hypothesis, however, is very good science that is playing a major role in the scientific paradigm shift now unfolding around us.

Old-paradigm science separated the living and the nonliving as mutually exclusive. We geologists studied rocks, while the biologists did microorganisms, plants, animals, fungi, and viruses. Now we are seeing that the line is arbitrary and in some ways false. Surely, I am alive and this rock in my hand is not. Yet my life depends upon this rock and others like it to produce and maintain the chemical constituents that form my body, the air I breathe, the water I drink. Without this rock, I cannot be. Gaia is made up of both the living and the nonliving joined in one complex overall organism.

The Gaia hypothesis emphasizes the ability of Earth's living and nonliving system to achieve a homeostatic balance that supports life. Yet we must not derive false comfort from Gaia's ability to heal. Within limits, the planet can respond, but misuse or abuse of Earth's systems could lead to global disasters. Gaia is something like the human body. A cut on the finger, a cold in the head, a pull in a muscle poses only a small challenge to our inherent healing mechanisms. But a burst of assault-rifle fire in the chest is likely to be too much, even if the victim has access to a well-equipped trauma center. From some wounds the body simply cannot recover.

No catastrophe in the past that we know of eliminated every living thing. Chicxulub killed all the dinosaurs, yet smaller reptiles and many mammals made it, eventually evolving into the range of animals we see about us today. Likewise, a comet streaking into Earth next month or year would not kill everything, unless it was unimaginably large. Some species would survive and, over the next millions and tens of millions of years, take advantage of new evolutionary opportunities and Gaia's eventual return to homeostasis. Life, though changed, would almost certainly go on.

Yet much more is at stake than the simple hand-to-mouth of biological survival. We civilized humans live in an exquisitely complex, highly interwoven web of relationships with the nonliving world, other living species, and each other. Much of what makes our lives interesting and worthwhile draws from this complexity. Were it shattered, our bodies might survive, but the quality of our way of life would be impoverished. Take the end of the Bronze Age as an example. Despite what must have been a heavy death toll, human life in the eastern Mediterranean was nowhere near snuffed out by the annual flurry of incoming comets. Yet there followed several centuries of a dark age in which the major accomplishments of the prior millennium were lost. Likewise, we as yet do not know what became of the civilization that carved the original Great Sphinx of Giza, and a similarly dark age divided the time of the Sphinx-builders from the period of the pyramids in the Old Kingdom. Surely, we ourselves wish to avoid the same fate.

Yet the odds for such an outcome may be heightened because of what we humans are doing, often inadvertently and unintentionally, to the planet. Recovery from a catastrophe, or any other disaster, requires a resilient Gaian system, one with the capacity to move back to its homeostatic state after a major disturbance. Yet without realizing what we are doing, we are changing the physical systems of the Earth in two primary ways that will make the consequences of an asteroid impact even worse and may well compromise the planet's—and our—ability to recover.

Patching the Hole in the Sky

One of these ways is depletion of the layer of ozone (O_3) naturally found in the upper level of the atmosphere known as the stratosphere and most strongly concentrated at an altitude of twelve to sixteen miles. Much of the time, oxygen is in the diatomic state, or O_2. When diatomic oxygen is struck by high-energy ultraviolet radiation, usually in the stratosphere, the diatomic molecule splits into two separate oxygen atoms. Each of the oxygen atoms can then combine with another diatomic oxygen to form ozone ($O_2 + O \leftrightarrow O_3$). Other reactions in the stratosphere remove ozone. If an ozone molecule absorbs ultraviolet radiation, it splits into diatomic oxygen and a single oxygen atom. The oxygen atom can then either recombine with a diatomic oxygen molecule to form ozone once again, combine with another atomic oxygen to form diatomic oxygen, or combine with some other substance in the stratosphere. Before humans began interfering with the atmosphere, ozone production equaled ozone destruction, so that the stratosphere always contained a small amount of ozone—a classic example of one of Gaia's many homeostatic mechanisms.

Though small in quantity, stratospheric ozone is essential to the preservation of current forms of life on the Earth because it acts as a shield to keep out the biologically dangerous form of ultraviolet radiation known as UV-B. When an ozone molecule is hit by UV-B and broken into diatomic oxygen and a single oxygen atom, the radiation's energy is absorbed and heat given off. This reaction keeps most of the UV-B radiation from reaching Earth's surface and, by giving off heat, contributes to relatively stable climatic conditions on and near the ground.

As is widely known, the concentration of stratospheric ozone has been declining as a result of human interference, largely through our release of enormous quantities of ozone-destroying substances into the atmosphere. The most important of these is a class of chemicals known as chlorofluorocarbons (CFCs) that, until recently, were widely used in refrigerators and freezers, automobile air conditioners, and aerosol cans; in the production of Styrofoam and other plastics; in insulators; and in solvents and cleaning agents used in the electronics industry. CFCs are almost inert in the atmospheric layer that covers the immediate surface of Earth, and eventually air currents carry them up into the stratosphere. There, ultraviolet radiation breaks down the CFC molecules, releasing atomic chlorine. The chlorine then reacts with an ozone molecule, removing one oxygen atom from the ozone, which then combines with the chlorine to form chlorine monoxide. The remainder of the ozone molecule is converted into diatomic oxygen. Since chlorine monoxide is unstable, it commonly reacts with an oxygen atom, which releases the chlorine atom to destroy yet another ozone molecule. Once released from a CFC, the single chlorine atom acts as a catalyst that can destroy tens of thousands of ozone molecules.

After this process became understood in the 1970s, a hue and cry was raised over the effects of increased UV-B radiation resulting from ozone depletion. There was good reason to be concerned. A small amount of UV radiation is necessary for well-being, because it promotes the synthesis of vitamin D in humans and acts as a germicide to control disease-causing microorganisms. Yet an increase in UV-B radiation at Earth's surface can have numerous adverse effects on health, primarily by causing skin cancer, including dangerous malignant melanomas, and cataracts, which impair or destroy vision.

Ultimately, the widespread ecosystem damage caused by increased UV-B may be more deleterious than its health effects. Abnormally high levels of UV-B radiation inhibit photosynthesis, metabolism, and growth in a number of plants, including important food crops like soybeans, potatoes, and wheat, as well as destroying cells and causing mutations. Since many tree species are particularly sensitive to UV levels, increasing amounts of UV may result in a major decline in forest productivity. Elevated levels of UV radiation affect insects, which are key to many terrestrial ecosystems. Numerous aquatic and marine plants and animals are extremely sensitive to UV levels, a particular concern in that UV-B can penetrate several yards of water. Phytoplankton—the plants and algae that form the basis of many food chains—and fish, crab, and shrimp larvae appear to be especially susceptible. Thus ozone depletion and an increase in UV-B radiation could seriously deplete our food supply. They add, too, to the stress put on the Gaian system, by weakening the living portion of the Earth organism.

The situation would only get worse if—or, more accurately, when—a large piece of cosmic rock or ice falls out of the sky and explodes. Nitrous oxide, another gas that attacks stratospheric ozone, is formed in the atmosphere in the extreme temperatures of a bolide blast. Research by Richard P. Turco and colleagues suggests that the Tunguska collision created so much nitrous oxide that up to 30 percent of the planet's stratospheric ozone was stripped away. In the early years of this century, before modern industry was producing and releasing CFCs and other ozone-destroying gases, the atmosphere's homeostatic mechanisms replaced the destroyed ozone in relatively short order. But think what would happen if another bolide the size of Tunguska appeared now, with the ozone layer already compromised. The UV-B effect likely to result from an impact would be increased, with deleterious effects on human health and the ability of many organisms, both plant and animal, to survive.

Hothouse Earth

The same point holds true for global warming, a phenomenon that is much more complex than ozone depletion. The threat to the ozone layer is relatively simple: One important environmental constituent is affected directly by several human-made industrial products. The link from cause to effect is straightforward, and the solution apparent. Global warming poses a vastly more complex body of issues and a potentially greater threat to civilization, one that would become vastly more complicated—and more dangerous—in the event of a bolide impact.

As practically everyone who picks up a newspaper or turns on a television knows by now, global warming results from what is known as the greenhouse effect, a natural homeostatic mechanism that humans have unwittingly tampered with. Because we have been filling the atmosphere with increased amounts of so-called greenhouse gases—particularly carbon dioxide, CFCs, and methane (natural gas), all of which tend to retain solar radiation and slow its passage into space—Earth has been warming up. In three of the past eight years—1990, 1995, and 1997—mean average temperatures were higher than in any other years since at least A.D. 1400. Global average temperature has risen a little less than one degree Fahrenheit in the past two decades.

Such a rise in global temperature in twenty years sounds less than dramatic, yet it is actually very rapid indeed. The warming trend that ended the most recent ice age came quickly by natural standards, yet this warming occurred at an average rate of approximately 1.8 degrees Fahrenheit every 500 or 600 years (although, as we saw in chapter 5, a Greenland ice core records a fourteen-degree jump in only fifteen years during the tenth millennium B.C.). The current global warming rate is some ten times or more faster than the natural average annual global increase in temperature at the end of the last ice age—a rise that is nothing short of ominous. And it's not going to stop anytime soon. In 1992 a group of about 200 scientists working under the auspices of the United Nations Intergovernmental Panel on Climate Change concluded that a rise in global mean temperature of 4.5 degrees Fahrenheit can be expected by 2100.

As Earth heats up, the warming will be unequally distributed, with higher latitudes generally warming up more, relative to the lower latitudes; the least change will occur in the equatorial regions. Climatic patterns will shift, and in some places local weather conditions will become much more violent. Cold air currents may be displaced such that, ironically, regions that are currently relatively warm may experience unusual cold snaps and abnormal winter storms. Changes in rainfall patterns might turn the American Midwest into a dust bowl, as the rain that has fallen in this region moves north into the Canadian prairies. Rainfall could also shift from one season to another. Some agricultural regions may receive more rain on average than at present, but the bulk of it will come during the winter months, when it is of little use for crops. As a result, an increase of only 3.6 degrees Fahrenheit might cause a decline in grain yields of up to 17 percent in North America and Europe, a loss of food this hungry world can ill afford.

The paths of ocean currents will also shift. Currently, for instance, cold water sinks in the Arctic Ocean and then travels southward along the bottom of the seas toward the equator. In the tropics the water is warmed and then moves north once again as a surface current, forming what can be thought of as a giant convection cell. With global warming, the northern polar regions may heat up enough to disrupt the convection cell; adequate quantities of cold water will no longer sink to the bottom to drive the surface currents back to the pole. The Gulf Stream in particular may slow down, stop flowing, or even change its direction. This could dramatically change local weather and climate patterns. England's weather, for instance, is heavily influenced by the warm waters of the Gulf Stream passing by its shores. If global warming disrupts the Gulf Stream, British winters may become much colder and harsher. Also, disrupting deep-water currents may affect the circulation of nutrients in the oceans, exacting a heavy toll on the extremely delicate ecosystems of the oceans.

As global warming continues, sea level will certainly rise. Water expands as it warms, and the possible melting of the polar ice caps will further swell the oceans. The 1992 United Nations Intergovernmental Panel on Climate Change (IPCC) report estimates that the rise in sea level will amount to about nineteen inches by the year 2100. Like the increase in average temperature, that number sounds small, but its effects will be devastating. Approximately one-third of the world's population lives within thirty-seven miles of the sea, often at elevations so close to sea level that a rise in the oceans would force hundreds of millions of people to higher ground. The IPCC estimated that a mere three-foot rise in global sea level, a distinct possibility within the next few centuries given current trends in global warming, would flood almost 250,000 miles of coastlines around the world.

Species other than humans would be affected. As climatic zones migrate away from the equator, trees will find themselves living in environments for which they are not adapted. Very small changes in average temperatures could spell death for large tracts of forest. One analysis suggests that if the concentration of carbon dioxide in the atmosphere doubles, the ranges of such trees as the eastern hemlock, yellow birch, beech, and sugar maple would have to migrate 300 to 600 miles north. If the warming occurs rapidly, as is expected, many of the trees will simply die. A large number of plant and animal species require very narrow ranges of temperature and moisture. As greenhouse warming modifies their habitats, they will find it difficult to adapt or migrate. Many will succumb and become extinct.

This already gloomy picture grows far gloomier when a major bolide hits Earth. Global warming, like damage to the ozone layer, weakens the Gaian system. Add the widespread devastation and global effects of cosmic collision, and a big problem becomes a catastrophe.

Consider what is likely to happen if a bolide two-thirds of a mile or more in diameter strikes the ocean. The most dangerous short-term effect of a marine impact is the series of tsunamis pulsating out from the point of the collision. With sea level rising from global warming, the impact waves will be more devastating, sweeping farther into the continental landmasses, killing populations crowded into coastal urban centers, subjecting even more land to saltwater contamination.

An oceanic impact would also disturb Earth's climate. One of the effects of global warming is a rise in atmospheric moisture. After all, with everything hotter, water evaporates more rapidly, creating conditions that are wetter overall and more likely to produce severe storms. The collision of a bolide into the ocean would multiply this effect. The impact would vaporize an immense quantity of sea water, which would add to the load of moisture in the atmosphere. Colossal rainstorms of the sort that prompted the stories of Noah and Utnapishtim would sweep across the continents and create devastating floods. If the bolide cracks the ocean floor and hot magma flows out into the sea, evaporating even more water, atmospheric moisture would continue at high levels and the planet could turn wet and warm. The effect, in essence, would be a kick-start for catastrophic global warming, with further rises in sea level, more and more storms, greater shifts in rainfall patterns, and wholesale changes in ocean currents. The death toll among humans would be enormous, and our civilization could well collapse under the strain of such changes.

A continental impact would be just as catastrophic, but in the opposite climatic direction: It could start the next ice age. As we saw in chapter 6, the dust veil created by a collision into a landmass augments the greenhouse effect by trapping heat that otherwise would have escaped into space. Evaporation from the oceans into the water-heavy atmosphere would increase in this hotter atmosphere. With the planet moving toward temperature equilibrium—cooling the lower-latitude oceans while warming the higher-latitude ones—violent storms would arise, which would be all the more powerful because of the increased amount of water vapor in the warmed atmosphere. Even though the high latitudes would rise in temperature, they still would remain more than cold enough for the colossal precipitation of the raging storms to accumulate as snow and turn into ice, laying the foundation for new glaciers. Within the three years it would take for the dust veil to fall back to Earth, enough snow and ice would have built up that it would reflect a larger proportion of the sun's energy back into space and cool the planet. The snowfields and glaciers that had begun in that first three years could continue to grow until another ice age was under way.

Perhaps this is the ultimate irony of human environmental tampering: By warming Gaia too rapidly, we have set the stage for catastrophe by ice.

Three Efforts for the Future

The new emerging paradigm, particularly the Gaia hypothesis, reminds us that we humans depend on the plants, animals, and physiochemical systems of Earth for our survival. For 4 billion years the planet's biological and nonbiological systems have survived a wide variety of onslaughts, including true polar wander, bolide impacts, ice ages, earthquakes, and widespread vulcanism, and still somehow recovered. Geological history shows the system to be resilient, but recent human actions are stretching this resilience to its limits.

To build responsibly for the human and planetary future, we

must do three things: preserve the ozone layer, counter global warming, and protect the planet against colliding asteroids and comets.

Ozone: What's Being Done?

It is possible that the beginning of the end of the ozone crisis came in June 1990, when ninety-three countries meeting in London agreed to end the production of CFCs and various other ozone-destroying chemicals by the year 2000 (with the exception of certain developing countries, which have until 2010 to stop CFC production). An international fund of more than $200 million was also established to assist the developing countries in switching to CFC substitutes, which often cost more than the CFCs they replace. The latter provision recognized an important aspect of the global issue: Most of the CFCs have been dumped into the atmosphere by the rich, industrialized countries, who then expect rich and poor alike to help solve the problem. The developing countries, however, want the refrigerators, air conditioners, electronics, and other conveniences common in the West. Only if the developed world contributes to the developing world can global cooperation on the ozone issue be achieved.

Additionally, a number of nations met again in Copenhagen in 1992 to update an earlier agreement signed in Montreal. The proposed phase-out date for all CFCs was moved up to January 1, 1996, with a ten-year grace period for the developing countries.

Even with the complete phase-out of CFCs and related chlorine-bearing chemicals, however, stratospheric chlorine concentrations are expected to remain high well into the twenty-first century and beyond, as already released CFC molecules rise into the stratosphere, a journey that can take fifteen years. Some CFCs have life expectancies of between 75 and 110 years, and the chlorine catalyst in the stratosphere can also be quite long-lived. As a result, scientists estimate that the ozone layer will continue to be depleted until at least 2050, resulting in ozone losses of as high as 10 to 30 percent over the northern latitudes, where most of the world's population resides. During the latter half of the next century, stratospheric ozone may start to build up again, but measurable amounts of CFCs will continue to reside in the atmosphere well into the twenty-fourth century. Despite our best efforts, the ozone layer will continue to thin for several decades to come.

Furthermore, CFCs are not the only potential threat to the ozone layer. Serious concern is being raised over other classes of chemicals that may also be contributing to ozone thinning. Some researchers are convinced that methyl bromide, a widely used agricultural pesticide, poses the biggest threat of all. Some studies indicate that a molecule of methyl bromide can break down ozone about 40 to 50 times as quickly as a CFC molecule, and it has been suggested that methyl bromide may currently account for 10 percent of all stratospheric ozone damage. As a result of such allegations, the U.S. Environmental Protection Agency and agencies of the United Nations have suggested complete bans on the production and use of methyl bromide.

Even with the worldwide control of CFCs, stratospheric ozone depletion will remain a problem well into the future. Still, the ozone story provides the basis for cautious optimism, as scientific understanding prevails and provides the basis for prudent policy. And the nations of the Earth have displayed a willingness to cooperate in the pursuit of a global solution to a global problem. When we humans come together in this intelligent and cooperative spirit, Gaia is well served.

Global Warming: Can the Trend Be Stopped?

Even though global warming is a clear threat that needs to be countered by effective action, most governments have pursued a weak compromise strategy, opting more for show than substance. Following the 1992 Earth Summit sponsored by the United Nations in Rio de Janeiro, Brazil, more than 160 countries signed the United Nations Framework Convention on Climate Change (UNFCCC), which requires its signatories to control emissions of greenhouse gases at specified levels. According to many scientists, however, the convention does not go nearly far enough. A follow-up meeting in December 1997 in Kyoto, Japan, produced a new protocol requiring the signing nations to reduce six key greenhouse gas emissions by 5 percent below 1990 levels by 2012. That's a step forward—but is it enough?

The main ways to control the greenhouse effect are to cut down on emissions of carbon dioxide, methane, and CFCs into the atmosphere and to restore the natural resources, such as forests, that convert carbon dioxide to nongaseous forms of carbon. Interestingly, virtually all measures that curb greenhouse gas emissions have additional benefits as well. Reducing emissions of CFCs and related gases will protect the ozone layer as well as helping to mitigate the greenhouse effect. Carbon dioxide emissions can be readily controlled by increasing energy efficiency, decreasing global dependence on fossil fuels, and developing alternative power technologies such as solar and wind.

Another important source of greenhouse gases is the carbon dioxide released during deforestation. Saving Earth's forests will not only help lessen the greenhouse effect by removing carbon dioxide from the atmosphere and releasing diatomic oxygen, but will also contribute to the preservation of biodiversity, help protect soils, and soften climatic extremes.

Clearly, we are falling short of stopping global warming. The problem is more difficult than ozone depletion because its sources are more diverse and complex and because they cut more deeply to the basic industrial processes of our civilization. The wake-up call has come in, but direct, concerted action is still lacking. The potential for disaster—slow, inevitable, and ultimately catastrophic—remains.

Impacts: Can We Dodge Bullets in the Cosmic Shooting Gallery?

Although no asteroid or comet of sufficient size to do continental damage is likely to fall in my lifetime or in yours, some such collision is bound to happen sooner or later. Regular near misses serve as reminders of just how little separates us from catastrophe. In March 1989, an asteroid named 1989FC, with an explosive potential equivalent to 2,000 megatons of TNT—approximately 100,000 times more than the Hiroshima blast—came within seven hours of hitting Earth and setting off a major regional catastrophe. On December 8, 1992, the asteroid Toutatis missed Earth by a distance equal to twice the span between our planet and the moon. In astronomical terms, that was like ducking the bullet—or, given Toutatis's size, the big bomb. Toutatis is 2.5 miles across, large enough to set off a blast equal to 9 million megatons of TNT, an explosion sure to result in global disaster.

Fortunately, astronomers know of no comet or asteroid likely to collide with Earth any time in the next century. The phrase "know of" is crucial, however. According to various estimates, somewhere between 1,500 and 3,000 objects more than two-thirds of a mile wide move in orbits that intersect Earth's. Of these, however, we have detected only about a hundred. The rest still have to be discovered, mapped, and tracked. Given this gap in our knowledge, there exists the potential that an object, even a relatively large one, could get close to Earth before we became aware of its presence. Asteroid 1989FC, the one that passed within seven hours of Earth in 1989, was detected only a few days in advance of its closest approach. The threat may be greatest with spent comets, which are as black as chimney soot, give off no easy-to-spot tail, and fly at speeds typically twice that of asteroids—100,000 miles per hour versus 50,000 miles per hour.

The first step to be taken in preparing to meet the threat of catastrophe from space is finding all the near-Earth objects and determining which ones pose a risk of collision. Eugene Shoemaker, the co-discoverer of P/Shoemaker-Levy 9 and a lifelong student of impacts and their threat, proposed a project called Spaceguard to serve as a planetary surveillance system. Spaceguard would comprise six large telescopes located around the world and linked to a central data-processing facility for analysis, verification, and tracking. Shoemaker estimated that with this commitment of resources, about 75 percent of NEOs more than two-thirds of a mile in diameter could be detected in approximately twenty-five years. The cost was steep—$50 million in initial investment, $10 million in annual maintenance—yet, given the economic results of a major impact as well as the loss of life, the trade-off seemed small indeed.

Still, our planet's nations have not exactly rushed to embrace the Spaceguard idea and pick up the tab. Until recently, for example, the entire annual expenditure by the United States federal government for NEO detection was only $1 million. Now that amount has risen to $3 million, a substantial increase, yet still not enough to do the necessary work as quickly as it should be done.

Currently, five telescopes are working full- or part-time at NEO detection. The Spacewatch project, run by Tom Gehrels and Robert S. McMillan of the University of Arizona's Lunar and Planetary Laboratory, has been at the task the longest, since 1989. Near-Earth Asteroid Tracking (NEAT), a cooperative project between NASA/Jet Propulsion Laboratory and the United States Air Force, is now operating from an observatory on the Haleakala volcano on the island of Maui. The air force is also funding the Lincoln Near Earth Asteroid Research (LINEAR) project run by the Massachusetts Institute of Technology, which uses a telescope on the White Sands Proving Ground in Socorro, New Mexico. The Lowell Observatory Near-Earth Object Search (LONEOS) system, which came online in January 1998, is sited in Flagstaff, Arizona. France and Germany are collaborating in the OCA-DLR Asteroid Survey (ODAS), which combines the resources of the Observatoire de la Côte d'Azur in Nice, France, with those of the Institute of Planetary Exploration in Berlin-Adlershof, Germany. The Minor Planet Center (MPC) at the Smithsonian Institution in Washington, D.C., serves as the central clearinghouse for the data gathered in the telescopes' systematic searching of the skies.

In addition to finding the asteroids and comets that pose the greatest threat, we need to know more about what they're made of and how they're put together. This kind of information is critical to knowing what we can do to stop, destroy, or deflect an NEO heading Earth's way. A few satellites have performed flybys of NEOs. An even more advanced mission, the Near Earth Asteroid Rendezvous (NEAR) satellite launched in February 1996, has already done a flyby of asteroid Mathilde in June 1997 and is scheduled to explore the asteroid Eros beginning in January 1999. Providing the first quantitative and comprehensive measurements of an asteroid's composition and structure, NEAR will aim to characterize Eros's physical and geological properties, such as the elements and minerals it contains and its density, shape, spin state, and internal structure. Flyby missions have also been planned by the European Space Agency, the German Center of Applied Space Technology, and the Japanese Institute of Astronomical and Space Science.

The importance of this kind of research is underscored by an old joke physicists like to tell. The joke, which has to do with physics' need for mathematical abstraction, ends with the punchline, "We assume a spherical chicken." You and I know, of course, that chickens aren't spherical. Chickens are in fact uniquely chicken-shaped, yet the mathematics of the chicken shape are difficult to handle. Calculations become much easier if we assume, as physicists do, that chickens are perfect spheres. The result is elegant mathematics, with an overriding air of unreality.

We've been doing much the same with asteroids and comets. Rough calculations of our ability to deflect or destroy an NEO on a flight path into Earth rest on the assumption of a spherical object of homogeneous composition. That assumption is every bit as unreal as the spherical chicken. A recent computer-simulation study by a research team headed by E. Asphaug showed that the results of a collision, whether with another asteroid or a nuclear weapon, has much to do with the asteroid's physical structure. Voids, fractures, and faults, for example, dampen the shock wave and protect the farthest points of the object, while the zone right at the point of collision is pulverized. A large, complex comet, for example, formed of rock and tar bound by ice, might be only partially destroyed, with the remainder of its bulk fragmenting into a number of pieces that plummet into Earth like P/Shoemaker-Levy 9's fiery parade. This kind of study makes it clear that we need to know what we are attacking before we set out to attack it.

For the next fifty years at least—a period in which, fortunately, the impact threat to Earth appears to be small—the only technology capable of destroying or deflecting an asteroid or comet is the nuclear weapon. Stockpiling and deploying nuclear weapons against asteroids or comets does, however, pose enormous problems. To begin with, nuclear weapons of the sort that would be used in space have to be tested under the conditions of use, and such testing is illegal. The international Treaty Banning Nuclear Weapons Tests in the Atmosphere, in Outer Space, and Underwater, which went into effect in 1963 and was one of the first steps toward world disarmament, specifically forbids the very kind of testing that would be necessary. In addition, the Treaty on Principles Governing the Activities of States in the Exploration and Use of Outer Space, Including the Moon and Other Celestial Bodies, which has been part of international law since 1967, prohibits the deployment or use of any nuclear device in orbit or on the moon. The late Carl Sagan was deeply disturbed that the rush to develop a nuclear weapons system to defend Earth against incoming objects might undermine legal agreements that have contributed to the slow easing of nuclear tensions on Earth and the first tentative steps toward destroying the Cold War's accumulated nuclear arsenals. And the very fact of possessing and storing nuclear weapons raises the specter of theft by terrorist, attack by rogue nation, or detonation by accident.

Since Earth has an apparent breathing spell until the next

likely swarm of dangerous incoming bolides, we should devote ourselves to developing technologies that are less threatening to our own survival than nuclear weapons, yet have the capability of destroying or deflecting a cosmic object headed Earth's way. Various technologies have been proposed—futuristic Buck Rogers–type tractor beams, gravitational deflection devices, and chemical weapons that consume the object, much like salt on ice or acid on a rock. Research on these possibilities needs to begin, sooner rather than later.

The impact threat is a global fact of life, and it must be met with a global response, one that does away with nationalistic maneuvering. Even the United States Air Force recognizes this. The military task force that developed the basic outline for what it calls a planetary defense system (PDS) maintains that the system must be under international control through the United Nations if it is to be credible.

This necessary cooperation might be one of the major benefits of developing PDS. Creating the system would probably lead to new technologies that would prove useful in nondefensive uses. Even more important, the nations of the world would learn again that, when it comes to the big problems, we all have to learn how to get along. The sense of global community created in building and maintaining PDS could not only save the planet from major damage by an unwelcome space visitor, but also, by laying a new foundation for cooperation, help preserve us against the insane violence we have delivered upon each other since the beginning of history.

Gaia's Ancient Face

In *Gaia,* James Lovelock writes, "To my mind the outstanding spin-off from space research is not new technology. The real bonus has been that for the first time in human history we have had a chance to look at the Earth from space, and the information gained from seeing from the outside our azure-green planet in all its global beauty has given rise to a whole new set of questions and answers."

Much the same can be said of the study of the ancient past, which is showing us clearly how catastrophe has shaped our planet and our civilization. Some people think the world of long ago contains a body of great secrets, like some treasure trove of insights that will unlock the universe's hidden doors, or new, undiscovered technologies that will free us from drudgery. I doubt that such is the case. The gifts of the ancient world go much deeper.

When I first went to Egypt to study the Great Sphinx of Giza, I journeyed as a scientist must, with an open, blank mind not given to preconceptions. Because of that necessary professional standpoint, the Sphinx gave me answers I didn't expect and revealed things I didn't think were possible.

Yet the Sphinx gave me more than data. As I spent time with that great ancient monument, I realized the sense of mystery and awe with which these people of long ago approached their world. Looking at the universe, with its cosmic rhythms, untold beauty, and great dangers, they understood themselves as part of something bigger than they themselves were. They knew their place in the order of things.

We need to recover that sense of the world. As we come together to understand and work against global warming, ozone depletion, and the impact threat, we too discover some corner of the mystery we live in, the grandeur of the life we contain and the universe we occupy. In waking up to the new paradigm, we recover something long lost, something very old, in ourselves.

Sources

Chapter 1. The Changing of the Paradigm

Clube, Victor, and Bill Napier. *The Cosmic Winter.* Oxford: Basil Blackwell, 1990.

Cramer, John G. "The Pump of Evolution." http://mist.npl.washington.edu/av/altvw11.html. 29 December 1997.

Eldredge, Niles, and Stephen Jay Gould. "Punctuated Equilibria: An Alternative to Phyletic Gradualism." In *Models in Paleobiology,* edited by T. J. M. Schopf, 82–115. San Francisco: Freeman, Cooper, 1972.

Kuhn, Thomas S. *The Structure of Scientific Revolutions.* Chicago: University of Chicago, 1962, revised 1970.

Palmer, Trevor. "The Fall and Rise of Catastrophism." http://euler.ntu.ac.uk/lsstaff/fallc.htm. 15 December 1997.

Schoch, Robert M. *Stratigraphy: Principles and Methods.* New York: Van Nostrand Reinhold, 1989.

Chapter 2. A Shape with Lion Body and the Head of a Man

Dobecki, Thomas L., and Robert M. Schoch. "Seismic Investigations in the Vicinity of the Great Sphinx of Giza, Egypt." *Geoarchaeology* 7, no. 6, 527–644 (1992).

Hassan, Selim. *The Sphinx: Its History in the Light of Recent Excavations.* Cairo: Government Press, 1949.

Lehner, Mark. *The Development of the Giza Necropolis: The Khufu*

Project. Mitteilungen des Deutschen Archäologischen Instituts, Cairo, vol. 41, 109–143 (1985).

Lehner, Mark. (Interview by A. R. Smith.) "The Search for Ra Ta." *Venture Inward* (magazine of the Association for Research and Enlightenment and The Edgar Cayce Foundation), January/February 1985, 6–11, 47; March/April 1985, 6–11.

——. "Computer Rebuilds the Ancient Sphinx." *National Geographic,* April 1991, 32–39.

——. "Reconstructing the Sphinx." *Cambridge Archaeological Journal* 2 (1), 3–26 (1992).

Schoch, Robert M. (Interview by A. R. Smith.) "The Sphinx: Older by Half?" *Venture Inward* (magazine of the Association for Research and Enlightenment and The Edgar Cayce Foundation), January/February 1992, 14–17, 48–49.

——. "Scholars Debate Age of the Great Sphinx" (letter). *The Chronicle of Higher Education,* 15 January 1992, B5.

——. "Redating the Great Sphinx of Giza." *KMT, A Modern Journal of Ancient Egypt* 3, no. 2 (summer 1992), 52–59, 66–70.

——. "A Modern Riddle of the Sphinx." *OMNI* 14, no. 11 (August 1992), 46–48, 68–69.

——. "Reconsidering the Sphinx." *OMNI* 15, no. 6 (April 1993), 31.

——. "Dating the Sphinx." *Condé Nast Traveler* 28, no. 2 (February 1993), 103.

Schwaller de Lubicz, René Aor. *Sacred Science: The King of Pharaonic Theocracy.* Translated by A. and G. VandenBroeck. New York: Inner Traditions International, 1982.

West, John Anthony. *Serpent in the Sky: The High Wisdom of Ancient Egypt.* Wheaton, Ill.: Quest Books/The Theosophical Publishing House, 1993.

Chapter 3. Ancient Origin: Civilization's Rescheduled Beginning

Bauval, Robert, and Adrian Gilbert. *The Orion Mystery: A Revolutionary New Interpretation of the Ancient Enigma.* New York: Crown Trade Paperbacks, 1994.

Campbell, Joseph. *The Mythic Image.* Bollingen Series C. Princeton, N.J.: Princeton University Press, 1974.

Edge, Frank. "Aurochs in the Sky: Dancing with the Summer Moon. A Celestial Interpretation of the Hall of Bulls from the Cave of Lascaux." Unpublished. December 1995.

——. "A Celestial Interpretation of the Hall of Bulls from the Cave of Lascaux." http://www.jse.com/absource.html. 19 December 1997.

——. "Les Aurochs de Lascaux dansant avec la Lune D'Été." *Kadath: Chroniques des Civilisations Disparues* 90, 20–34 (spring-summer 1998).

Graves, Robert. *The White Goddess: A Historical Grammar of Poetic Myth.* New York: Farrar, Straus & Giroux, amended and enlarged edition, 1966.

Hancock, Graham, and Robert Bauval. *The Message of the Sphinx: A Quest for the Hidden Legacy of Mankind.* New York: Three Rivers Press, 1996.

Marshack, Alexander. *The Roots of Civilization: The Cognitive Beginnings of Man's First Art, Symbol, and Notations.* New York: McGraw-Hill, 1972.

NBC News. "Linguistics, Anthropology Hint at a Complex Ancient Odyssey." http://www.nbcnews.com/news/144348.asp. 15 April 1998.

Office of Public Relations, University of Colorado. "Oldest Astronomical Megalith Alignment Discovered in Southern Egypt by Science Team." http://www.colorado.edu/PublicRelations/NewsReleases/1998/Oldest_Astronomical_Megalith_A.html. 29 May 1998.

Phillips, Helen. "Archaeology: Desert Astronomers." http://www.nature.com/Nature2/serve?SID=46204898&CAT=Corner&PG=Update/update590.html. 29 May 1998.

Saleh, Mohamed, and Hourig Sourouzian. *Official Catalogue: The Egyptian Museum.* J. Liepe, photographer. Mainz: Verlag Philipp von Zabern, 1987.

Santillana, Giorgio de, and Hertha von Dechend. *Hamlet's Mill:*

An Essay on Myth and the Frame of Time. Boston: Gambit, 1969.

Settegast, Mary. *Plato Prehistorian: 10,000 to 5000 B.C. in Myth and Archaeology.* Cambridge, Mass.: Rotenberg Press, 1986.

Vermeersch, P. M., E. Paulissen, G. Gijselings, M. Otte, A. Thoma, P. van Peer, and R. Lauwers. "33,000-yr Old Chert Mining Site and Related *Homo* in the Egyptian Nile Valley." *Nature* 309, 342–44 (1984).

Wilford, John Noble. "Excavation in Chile Pushes Back Date of Human Habitation of Americas." http://www.latinolink.com/news/news97/0210NEXC.HTM. 15 April 1998.

Chapter 4. Looking for the Lost Cities

Anonymous. "Lemurian Paradise." http://www.crystalinks.com/lemeura.html. 19 June 1998.

Bauval, Robert, and Graham Hancock. "The Mysterious Structures That May Upstage NASA's Evidence of Martian Life." http://www.planetarymysteries.com/sphinxmars.html. 12 March 1998. Reprinted from the *Daily Mail* (London), 17–19 August 1996.

Carlotto, Mark J. *The Martian Enigmas: A Closer Look.* Berkeley: North Atlantic Books, 1991.

Chandler, David L. "Mars Life Theory Gains Momentum." http://www.boston.com/dailyglobe/globehtml/080/Mars_life_theory_gains_momentum.htm. 21 March 1998.

Chang, Kenneth. "The Martian Rohrschach Test." http://www.abcnews.com/sections/science/DailyNews/marsface980407.html. 10 April 1998.

Däniken, Erich von. *Chariots of the Gods?: Unsolved Mysteries of the Past.* Translated by Michael Heron. New York: Putnam, 1970.

Donnelly, Ignatius. *Atlantis: The Antediluvian World.* New York: Harper and Brothers, 1882.

Flem-Ath, Rand. "Atlantis and the Earth's Shifting Crust." Return to the Source Symposium, University of Delaware, 28 September 1996. http://www.netfeed.com/pstevens/flem-ath.html. 1 January 1998.

Flem-Ath, Rand, and Rose Flem-Ath. *When the Sky Fell: In Search of Atlantis.* New York: St. Martin's Press, 1995.

Hancock, Graham. *Fingerprints of the Gods: The Evidence of Earth's Lost Civilization.* New York: Crown Publishers, 1995.

Hapgood, Charles H. *Maps of the Ancient Sea Kings: Evidence of Advanced Civilization in the Ice Age.* Revised edition. New York: E. P. Dutton, 1979.

Heinrich, Paul. "The Mysterious Origins of Man: The Oronteus Finaeus Map of 1532." http://www.talkorigins.org/faqs/mom/oronteus.html. 19 June 1998.

Heinrich, Paul V. "Lemuria, a Scientific Frankenstein." http://www.m-m.org/jz/sphinxcc.html. 19 June 1998.

Hoye, Paul F., with Paul Lunde. "Piri Reis and the Hapgood Hypothesis." http://www.millersv.edu/~columbus/data/art/HOYE01.ART. 20 June 1998.

Kenyon, J. Douglas. "Atlantis in Antarctica?" http://members.aa.net/~mwm/atlantis/issue7ar7antarctica1.html. 13 March 1998.

Kukal, Zdeněk. *Atlantis: In the Light of Modern Research.* Translated by V. Zborilek and C. Emiliani. Amsterdam: Elsevier, 1984. Reprinted from *Earth Science Reviews* 21 (1984).

Livermore, Beth. "Antarctic Meltdown." *Popular Science,* February 1997, 38–43.

Lunde, Paul. "Piri Reis and the Columbian Theory." http://www.millersv.edu/~columbus/data/art/LUNDE01.ART. 19 June 1998.

——. "The Oronteus Finaeus Map." http://www.millersv.edu/~columbus/data/art/LUNDE02.ART. 19 June 1998.

McKay, D. S., E. K. Gibson Jr., K. L. Thomas-Keprta, H. Vali, C. S. Romanek, S. J. Clemett, X. D. F. Chillier, C. R. Maechling, and R. N. Zare, "Search for Past Life on Mars: Possible Relic Biogenic Activity in Martian Meteorite ALH84001." *Science* 273 (16 August 1996), 924–30.

National Science Foundation. "Why Is Antarctica So Cold? Scientists Pursue History of Antarctic Ice Sheet." http://www.geo.nsf.gov/adgeo/press/pr9811.htm. 8 June 1998.

Plato. *Timaeus and Critias.* http://www.activemind.com/Mysterious/Topics/Atlantis/timaeus_and_critias.html. Translated by Benjamin Jowett. 1 January 1998.

Santos, Arysio Nunes dos. "A Novel Theory on Atlantis." http://www.atlan.org/index.html. 14 March 1998.

Schiff, Joel L. "Martian Life—the Pros and Cons." http://www.meteor.co.nz/mars/html. 31 December 1997. Reprinted from *Meteorite,* November 1996.

Settegast, Mary. *Plato Prehistorian: 10,000 to 5000 B.C. in Myth and Archaeology.* Cambridge, Mass.: Rotenberg Press, 1986.

Sitchin, Zecharia. *The Twelfth Planet.* New York: Avon Books, 1976.

Chapter 5. Fire and Water

Alley, Richard B., and Michael L. Bender. "Greenland Ice Cores: Frozen in Time." *Scientific American,* February 1998, 80–85.

Anonymous. "Scientists Discover That 'Evolutionary Big Bang' May Have Been Caused by Earth Losing Its Balance Half a Billion Years Ago." http://www.caltech.edu/~media/lead/072497JLK.html. 2 May 1998.

Anonymous. "What Controls the Advance and Retreat of These Large Glaciers During the Four Long, Cool Periods?" http://www.museum.state.il.us/exhibits/ice_ages/why_glaciations1.html. 14 March 1998.

Barbiero, Flavio. "On the Possibility of Very Rapid Shifts of the Poles." http://www.unibg.it/dmsia.dynamics.poles.html. 15 December 1997. Also published in *Quaderni del Dipartimento di Matematica, Statistica, Informatica ed Applicazoni, Universita degli Studi di Bergamo, Italy,* 1997, no. 7, 1–20.

Charles, Arthur. "Evidence of Black Sea Beginning for Flood." http://churchnet.ucsm.ac.uk/news/files2/news167.htm. 27 June 1998. Reprinted from *The Independent,* 2 October 1997.

Collins, Truman. "The May 5, 2000, Planetary Alignment and Its Destructive Potential." http://www.teleport.com/~tcollins/conjunct.shtml. 12 April 1998.

Drews, Robert. *The End of the Bronze Age: Changes in Warfare and the Catastrophe Ca. 1200 B.C.* Princeton: Princeton University Press, 1993.

Emiliani, C., S. Gartner, B. Lidz, K. Eldridge, D. K. Elvey, T. C. Huang, J. J. Stipp, and M. F. Swanson, 1975. "Paleoclimatological Analysis of Late Quaternary Cores from the Northeastern Gulf of Mexico." *Science* 189, 1083–88.

Gee, Henry. "Earth: 1601 and All That." http://www.nature.com/Nature2/serve?SID=43105727&CAT=Corner&PG=Update/update623.html. 19 June 1998.

Gould, Stephen Jay. *Wonderful Life: The Burgess Shale and the Nature of History.* New York: W. W. Norton, 1990.

Graves, Robert. *The Greek Myths.* Revised edition, 2 volumes. New York: Penguin, 1960.

Hapgood, Charles H. *The Path of the Pole.* Philadelphia: Chilton Books, 1970.

Hartmann, Dennis L. "Our Changing Climate." http://eos.atmos.washington.edu/~dennis/OCC_Final_961216.html. 25 January 1998

Isaak, Mark. "Flood Stories from Around the World." http://www.talkorigins.org/faqs/flood-myths.html. 27 June 1998.

Kenyon, J. Douglas. "Richard Noone: The Ancient Past Warms Him Up, But the Future Makes Him Shiver." *Atlantis Rising* 14 (1998), 25–27.

Kettler, Al. "Earth-Changing Drama...in a Geologic Heartbeat." http://www.uidaho.edu/igs/iafi/iafidesc.html. 19 June 1998. First appeared in *Smithsonian,* April 1995, 50.

Kirschvink, Joseph L., Robert L. Ripperdan, and David A. Evans. "Evidence for a Large-Scale Reorganization of Early Cambrian Continental Masses by Inertial Interchange True Polar Wander." *Science* 277 (June 1997), 541–45. See also http://www.gps.caltech.edu/~devans/iitpw/science.html. 2 May 1998.

Margulis, Lynn, and Karlene V. Schwartz. *Five Kingdoms: An Illustrated Guide to the Phyla of Life on Earth.* Second edition. New York: W. H. Freeman, 1988.

Martini, Kirk. "Volcanic Phenomena at Pompeii." http://urban.arch.virginia.edu/struct/pompeii/volcanic.html. 6 July 1998.

Monson, Brian. "Planetary Conjunctions." http://www.physics.utulsa.edu/astronomy/Conjunctions.html. 12 April 1998.

Noone, Richard W. *5/5/2000: Ice: The Ultimate Disaster.* New York: Three Rivers Press, 1982.

Ochert, Ayala. "As the World Turns." http://www.nature.com/Nature2/serve?SID=18415757&CAT=Corner&PG=Update/update071.html. 10 May 1998.

Plait, Phil. "Bad Astronomy: Planetary Alignments Will Cause Earthquakes." http://smart.net~badastro/bad/planets.html. 12 April 1998.

Sandars, N. K., ed. and trans. *The Epic of Gilgamesh: An English Version with an Introduction.* Second revised edition. New York: Penguin Books, 1972.

Sawyer, Kathy. "Global Shift Tied to Evolution." http://www.washingtonpost.com/wp-srv/frompost/features/july97/continent.htm. 2 May 1998. Originally published in *The Washington Post,* 25 July 1997.

Steig, E. J., E. J. Brook, J. W. C. White, C. M. Sucher, M. L. Bender, S. J. Lehman, D. L. Morse, E. D. Waddington, and G. D. Clow. "Synchronous Climate Changes in Antarctica and the North Atlantic." *Science* 282 (2 October 1998), 92–95.

Strain, Mac. B. *The Earth's Shifting Axis: Clues to Nature's Unsolved Mysteries.* Shrewsbury, Mass.: ATL Press, 1997.

Swanson, Doug J. "'Perilous Beauty': Mount Rainier Considered the 'Most Dangerous Volcano in the United States.'" *San Francisco Examiner,* 12 July 1998.

Taylor, K. C., P. A. Mayewski, R. B. Alley, E. J. Brook, A. J. Gow,

P. M. Grootes, D. A. Meese, E. S. Saltzman, J. P. Severinghaus, M. S. Twickler, J. W. C. White, S. Whitlow, and G. A. Zielinksi. "The Holocene–Younger Dryas Transition Recorded at Summit, Greenland." *Science* 278 (31 October 1997), 825–27.

White, John. *Pole Shift: Scientific Predictions and Prophecies About the Ultimate Natural Disaster.* Virginia Beach, Va.: ARE Press, 1980. Epilogue added in 1990.

Wright, Herbert E., Jr. "Glacial Fluctuations, Sea-Level Changes, and Catastrophic Floods." In *Atlantis: Fact or Fiction?,* edited by Edwin S. Ramage. Bloomington: Indiana University Press, 1978, 161–74.

Chapter 6. Heaven's Rain of Rock and Ice

Akridge, Glen. "The Prehistoric Use of Meteorites in North America." http://www.meteor.co.nz/may96_2.html. 31 December 1997.

Anonymous. "Comet That Launched Noah's Ark." http://www.nando.net/newsroom/ntn/health/042296/health11_25615.html. 31 December 1997.

Anonymous. "Hazards Due to Near-Earth Objects." http://ds.dial.pipex.com/town/terrace/fr77/meet.html. 10 April 1998. Report of a meeting held at the Royal Greenwich Observatory in Cambridge, England, on 10 July 1997.

Blood, Michael L. "The Tektites of Tibet." http://www.meteor.co.nz/may96_3.html. 31 December 1997.

Boslough, Mark, and David Crawford. "Frequently Asked Questions About the Collision of Comet Shoemaker-Levy 9 with Jupiter: Post-Impact Questions and Answers." http://www.isc.tamu.edu/~astro/sl9/cometfaq2.html. 11 April 1998.

Bowdan, Scott. "A Comet's Fiery Dance at Jupiter." http://www.jpl.nasa.gov/sl9g1129.htm. 11 April 1998. Also appeared in *The Galileo Messenger,* May 1995.

British Broadcasting System. "It Came from the Skies." http://news.bbc.co.uk/hi/english/sci/tech/newsid%5F40000/40144.stm. 31 December 1997.

Campbell, Joseph. *The Masks of God: Occidental Mythology.* New York: Penguin Books, 1976.

Clube, Victor, and Bill Napier. *The Cosmic Winter.* Oxford, England: Basil Blackwell, 1990.

Desonie, Dana. "Threat from Space: The Science Behind the Movie *Deep Impact.*" http://www2.earthmag.com/earth/DeepImpact/asteroid.html. 2 May 1998.

Farinella, Paolo. "Chaotic Routes Between the Asteroid Belt and Earth." http://www.meteor.co.nz/may96_1.html. 31 December 1997.

Ferris, Timothy. "Is This the End?" http://matu1.math.auckland.ac.nz/~kingPreprints/book/future/comet. html. 1 January 1998. Contains excerpts from an article that appeared in *The New Yorker,* 27 January 1997.

Gallant, Roy A. "'The Sky Has Split Apart!: The Cosmic Mystery of the Century." http://www.galisteo.com/tunguska/docs/splitsky.html. 10 April 1998.

Geologic Division, Woods Hole Field Center. "The Chesapeake Bay Bolide: Modern Consequences of an Ancient Cataclysm." http://woodshole.er.usgs.gov/epubs/bolide/index.html. 31 December 1997.

Gurov, Eugene P. "Impact Craters of the Earth." http://www.meteor.co.nz/aug96_2.html. 31 December 1997.

Harris, A. "'Tunguska '96,' Bologna, Italy, 15–17 July 1996: A Meeting Report." http://ccf.arc.nasa.gov/sst/10-11-96.html. 15 December 1997.

Kobres, Bob. "Comets and the Bronze Age Collapse." http://abob.libs.uga.edu/bobk/bronze.html. 13 March 1998. Also published by the Society for Interdisciplinary Studies in *Chronology and Catastrophism Workshop* 1 (1992), 6–10.

Lewis, John S. *Rain of Iron and Ice: The Very Real Threat of Comet and Asteroid Bombardment.* Reading, Mass.: Addison-Wesley, 1996.

Muller, Richard. *Nemesis: The Death Star.* New York: Weidenfeld & Nicholson, 1988.

——. "Nemesis." http://muller.lbl.gov/lbl-nem.htm. 1 August 1998.

National Aeronautics and Space Administration. "Chain of Impact Craters Suggested by Spaceborne Radar Images." http://www.jpl.nasa.gov/s19/news80.html. 29 May 1998.

——. "Comet Shoemaker-Levy Background." http://www.jpl.nasa.gov/sl9/background.html. 11 April 1998.

National Science Foundation. "Scientific Mystery Remains in Australia: Scientist-Sleuths to Report on Latest Findings." http://www.nsf.gov/od/lpa/news/release/pr9741.htm. 31 December 1997.

Ovid. *Metamorphoses.* Translated by Rolfe Humphries. Bloomington: Indiana University Press, 1955.

Peiser, B. J. "Wave of New Publications Indicates Scientific Revolution Is Under Way." http://www.fisherman.se.zetatalk/theword/twword04s.htm. 24 December 1997.

Peiser, Benny. "Comets and Disaster in the Bronze Age." http://britac3.britac.ac.uk/cba/ba/ba30/ba30feat.html1#peiser. 12 June 1998. Reprinted from *British Archaeology* 30 (December 1997).

Persson, Henrik. Untitled. (1997 Greenland meteor impact.) http://www2.dk-online.dk/users/hpersson/meteor/press.htm. 31 December 1997.

Spedicato, Emilio. "Apollo Objects, Atlantis, and Other Tales: A Catastrophical Scenario for Discontinuities in Human History." http://www.unibg.it/dmsia/dynamics/apollo.html. 15 December 1997.

——. "Evidence of Tunguska-Type Impacts over the Pacific Basin Around the Year 1178 A.D." http://www.unibg.it/dmsia/dynamics/year.html. 13 March 1998.

Spray, John G., Simon P. Kelley, and David B. Rowley. "Evidence for a Late Triassic Multiple Impact Event on Earth." *Nature* 392 (1998), no. 6672, 171–73.

Steel, Duncan. *Rogue Asteroids and Doomsday Comets: The Search for the Million Megaton Menace that Threatens Life on Earth.* New York: John Wiley and Sons, 1995.

Stone, Richard. "The Last Great Impact on Earth." http://magazines.enews.com/magazines/discover/magtxt/090196–1.html. 31 December 1997. Also appeared in *Discover Magazine,* September 1996.

Strasenburgh Planetarium. "Collision with Asteroids and Comets." http://rmsc.org/html/planet/faqs/collisions.html. 1 January 1998.

University of Chicago. "Crater Chain on Two Continents Points to Impact from Fragmented Comet." http://www.flatoday.com/space/explore/stories/1998/040498c.htm. 29 May 1998.

Verschuur, Gerrit. *Impact! The Threat of Comets and Asteroids.* Oxford: Oxford University Press, 1996.

Chapter 7. Learning from the Past, Looking Toward the Future

Asphaug, E., S. J. Ostro, R. S. Hudson, D. J. Scheeres, and W. Benz. "Disruption of Kilometre-Sized Asteroids by Energetic Collisions." *Nature* 393 (4 June 1998), 437–41.

Calvin, William H. "The Great Climate Flip-flop." *The Atlantic Monthly,* January 1998, 47–64.

Chapman, Clark R. "The Asteroid/Comet Impact Hazard." http://k2.space.swri.edu/clark/ncar.html. 14 March 1998.

Farb, Michael. "James Lovelock's Gaia Hypothesis: Past, Present, and Future." http://www.slip.net/~farb/gaia_overview.htm. 17 March 1998.

Gee, Henry. "Climate: Catastrophic Collapse of Antarctic Ice Shelves." http://www.nature.com/Nature2/serve?SID=88508107&CAT=Corner&PG=Update/update500.html. 23 April 1998.

——. "Greenhouse Century." http://www.nature.com/Nature2/serve?SID=46204898&CAT=Corner&PG=Update/update570.html. 29 May 1998.

Lovelock, J. E. *Gaia: A New Look at Life on Earth.* Oxford: Oxford University Press, 1979.

Lovelock, James. "What Is Gaia?" http://www.ion.com.au/our planet/lovelock2.html. 17 March 1998.

McKinney, Michael L., and Robert M. Schoch. *Environmental Science: Systems and Solutions.* Web-enhanced edition. Sudbury, Mass.: Jones and Bartlett Publishers, 1998.

Rambler, Mitchell B., Lynn Margulis, and René Fester, eds. *Global Ecology: Towards a Science of the Biosphere.* Boston: Academic Press, 1989.

Schoch, Robert M. *Case Studies in Environmental Science.* Minneapolis/St. Paul: West Publishing Company, 1996.

Spedicato, Emilio. "Apollo Objects, Atlantis, and Other Tales: A Catastrophical Scenario for Discontinuities in Human History." http://www.unibg.it/dmsia/dynamics/apollo.html. 15 December 1997.

Stanley, Diana. "Dr. James Lovelock: Formulation of the Gaia Hypothesis." http://magna.com.au/~prfbrown/gaia_jim.html. 17 March 1998.

——. "Dr. Lynn Margulis: Microbiological Collaboration of the Gaia Hypothesis." http://magna.com.au/~prfbrown/gaia_lyn.html. 17 March 1998.

Toon, O. B., K. Zahnle, R. P. Turco, and C. Covey. "Environmental Perturbations Caused by Impacts." In *Hazards Due to Comets and Asteroids,* edited by T. Gehrels. Tucson: University of Arizona Press, 1994, 791–826.

Urias, John M., Iole M. DeAngelis, Donald A. Ahern, Jack S. Caszatt, George W. Fenimore III, and Michael J. Wadzinski. "Planetary Defense: Catastrophic Health Insurance for Planet Earth." http://www.au.af.mil/au/2025/volume3/chap16/v3c1 6-1.htm#Contents. 1 January 1998.

Vasilyev, N. V. "The Tunguska Meteorite Problem Today." http://www.galisteo.com/tunguska/docs/tmpt.html. 10 April 1998.

Index